Collins

Maths
Skills Builder
Transition from KS3 to GCSE

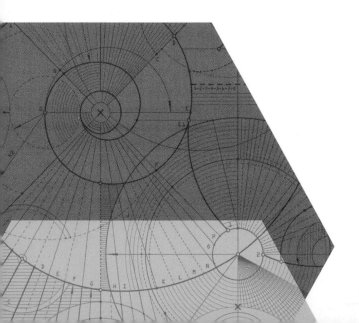

Chris Pearce

William Collins' dream of knowledge for all began with the publication of his first book in 1819. A self-educated mill worker, he not only enriched millions of lives, but also founded a flourishing publishing house. Today, staying true to this spirit, Collins books are packed with inspiration, innovation and practical expertise. They place you at the centre of a world of possibility and give you exactly what you need to explore it.

Collins. Freedom to teach.

Published by Collins
An imprint of HarperCollins*Publishers*
77–85 Fulham Palace Road
Hammersmith
London
W6 8JB

Browse the complete Collins catalogue at
www.collins.co.uk

© HarperCollins*Publishers* Limited 2014

10 9 8 7 6 5 4 3 2 1

ISBN-13 978-0-00-753780-8

Chris Pearce asserts his moral right to be identified as the author of this work.

British Library Cataloguing in Publication Data
A catalogue record for this publication is available from the British Library.

Commissioned by Katie Sergeant
Project managed by Elektra Media Ltd
Developed and copy-edited by Joan Miller
Proofread by Amanda Dickson
Edited by Helen Marsden
Illustrations by Ann Paganuzzi
Typeset by Jouve India Private Limited
Cover design by We Are Laura
Production by Robin Forrester

Printed and bound by L.E.G.O. S.p.A, Italy

Acknowledgements
The publishers wish to thank the following for permission to reproduce photographs. Every effort has been made to trace copyright holders and to obtain their permission for the use of copyright materials. The publishers will gladly receive any information enabling them to rectify any error or omission at the first opportunity.

(t = top, c = centre, b = bottom, r = right, l = left)

Cover images: bl Kiorafilms-JochenTeschke/Shutterstock Premier, tr Kachan Eduard/Shutterstock Premier, p 18 ipag/Shutterstock, p 20 Samuel Borges Photography/Shutterstock, p 29 Paul Hakimata Photography/Shutterstock, p 30t CreativeNature.nl/Shutterstock, p 30b Thomas Klee/Shutterstock.

Contents

Introduction

Welcome to Collins *Maths Skills Builder*. This book builds on the concepts and ideas you already know at Key Stage 3 and the maths skills you have gained. It will help you to build skills, ready for GCSE or an equivalent.

There will be new maths skills for you to learn and new topics to master at GCSE. But you will also need to remember and use the skills from Key Stage 3 and learn to apply those skills in a greater variety of ways.

You will learn to:

- apply your knowledge in more complex situations and choose appropriate methods
- communicate mathematically by giving explanations and reasons, using appropriate language
- make connections between different parts of maths.

This book is designed to give you an idea of what that means in practice. The questions are more than just straightforward applications of your maths skills. They should make you think a bit more and gain confidence in tackling problems that are different or unusual in some way.

The questions are divided into six sections that match the Key Stage 3 and Key Stage 4 Programme of Study:

- Number
- Algebra
- Ratio, percentage and rates of change
- Geometry and measures
- Probability
- Statistics.

There is also a seventh section of mixed questions that can be taken from any area of the Programme of Study.

The questions in each section are arranged in approximate order of difficulty. Harder questions at the end of each section are marked with an asterisk (*). Each section starts with a worked example. The sections are not all the same length: the length is approximately proportional to the amount of material in the Programme of Study.

Good luck! I hope you enjoy tackling these problems in preparation for GCSE.

1 Number

WORKED EXAMPLE

35% of £80 = ____% of £40

Work out the missing number.

SOLUTION

This is a question about percentages. You should know:

- how to work out a percentage of a quantity
- how to write one quantity as a percentage of another.

There are several ways to answer this question.

Method 1

35% of £80 = 0.35 × 80 = £28

So ___% of £40 = £28

The percentage is $\frac{28}{40} \times 100 = 0.7 \times 100 = 70\%$.

Method 2

35% of £80 = ____% of £40

Notice that the amount of money is divided by 2.

That means you multiply the percentage by 2 to get the same answer.

35% × 2 = 70%, as before.

You can use any method you like.

In this example it is easy to check your answer:
70% of £40 = 0.7 × 40 = £28 This is correct.

QUESTIONS

1. Zeta has these coins.

She can make different amounts with her coins.

For example, 17p = 10p + 5p + 2p.

a. Show how to make 68p with her coins.

b. What is the smallest whole number of pence that Zeta **cannot** make with her coins?

2. This thermometer shows the temperature of a freezer at 13:00.

Between 13:00 and 14:00 the temperature falls by 1.5 degrees.

Between 14:00 and 15:00 the temperature rises 3.7 degrees.

a. Work out the temperature at 15:00.

b. Work out the overall change in temeperature from 13:00 to 15:00.

3. Here are four number cards.

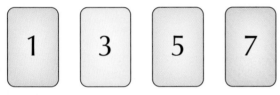

You can make two-digit numbers with pairs of these cards.

a. Write down a multiple of three you can make with two cards.

b. Work out the difference between the largest possible two-digit number and the smallest possible two-digit number.

4. Here are the frequencies for four FM radio stations, as listed in a magazine.

Look at this scale.

80   100

What station is the arrow pointing at?

5. Justine buys 4 bananas and 5 nectarines.

The bananas cost £1.80.

The total cost is £5.05.

a. What is the price of one banana?

b. How much does each nectarine cost?

6. a. Find a fraction to make this addition correct.

$$\frac{1}{10} + \underline{\quad} = \frac{3}{5}$$

b. Find a fraction to make this subtraction correct.

$$\underline{\quad} - \frac{1}{2} = \frac{1}{5}$$

7. a. Find a multiple of 5 that is also a multiple of 6.

b. Find a factor of 30 that is **not** a factor of 50.

c. Find the sum of all the factors of 28.

d. Find the product of all the factors of 10.

8. Work out the missing number.

$$3.5 + \underline{\quad} = 6.1 - 2.3$$

9. Look at each of the calculations. Put in brackets to make them correct.

a. $5 + 4 \times 8 - 3 = 25$

b. $5 + 4 \times 8 - 3 = 45$

c. $5 + 4 \times 8 - 3 = 69$

10. Look at these five numbers.

34   37   39   40   41

Which of these is **not** the sum of two square numbers? Give a reason for your answer.

11. a. Show that $\frac{3}{10}$ is more than $\frac{1}{4}$ but less than $\frac{1}{3}$.

b. Find a fraction that is more than $\frac{2}{3}$ but less than $\frac{3}{4}$.

12. Here are five numbers.

   −6   5   −4   3   −2

   a. Put the numbers in order, smallest first.

   b. What is the largest possible difference between two of these numbers?

   c. One of the numbers is the sum of the other four. Which one?

   d. What is the largest possible product of two of these numbers?

13. In 2010 an apartment sells for £167 500.

   In 2014 the same apartment sells for £212 400.

   a. Work out the increase in price.

   b. Tim says: 'The price increase is more than 25%.'

   Show that this is true.

14. $A \times B \times 7 = 1001$

   A and B are prime numbers.

   Work out the values of A and B.

15. Look at these numbers.

   31   35   37   39   41   45   49

   a. Which is a square number?

   b. Which is a multiple of 13?

   c. Which are the product of two prime numbers?

16. Look at this advertisement.

Special deal on sofa bed

Cash price
£800
or
Pay **just £149** and
10 monthly instalments of **£69**

Work out how much you save if you pay the cash price.

17. a. Show that $\frac{1}{4}$ of 0.6 is the same as $\frac{3}{4} - 0.6$.

   b. Is $\frac{1}{5}$ of 0.5 the same as $\frac{3}{5} - 0.5$? Give a reason for your answer.

18. Plant support sticks are available in three lengths.

Length of plant support stick	
Short	30 cm
Medium	60 cm
Long	80 cm

Support sticks are cut from rods 2.50 m long.

a. How many short sticks can be cut from one rod?

b. How many long sticks can be cut from one rod?

c. Is it possible to cut a 2.50 m rod into a mixture of sticks of different sizes without any waste? Give a reason for your answer.

19. This question is about the number 28.

a. Write 28 as the sum of two prime numbers.

b. Write 28 as the sum of three different prime numbers.

c. Write 28 as the sum of four different prime numbers.

d. Is it possible to write 28 as the sum of five different prime numbers? Give a reason for your answer.

20. Without using a calculator, work out the missing number in each of these calculations.

a. $3.4 \times \_\_\_ = 340$

b. $1000 \times \_\_\_ = 740$

c. $\_\_\_ \div 100 = 4.6 \times 10$

21. Plant food is sold in the form of granules.
This is the label on the container.

Dosage: Sprinkle 75 g around the base of each plant

One handful of granules is approximately 30 g.

a. How many handfuls are needed for each plant?

b. The jar contains one kilogram of granules. How many doses are there in one jar?

22. Bottles of water cost £0.90.

Packs of sandwiches cost £2.60.

Sean buys water and sandwiches and spends £11.40.

How many bottles of water does he buy?

23. A dance group hires a practice room for eight sessions.

The room costs £35 per session.

There are 13 people in the group.

They share the cost of room hire equally among them.

Work out how much each person pays.

24. Mel is paid £1350 per month.

She pays 15% of this in tax.

She pays one-third of it in rent.

Work out how much Mel has left each month after paying tax and rent.

25. Complete these statements.

  a. 44% of 120 kg = ____% of 60 kg

  b. 15% of £750 = 45% of £_____

*26. a. Put the numbers 5, 6, 7, 8 and 9 in the boxes so that every row has a total of 19.

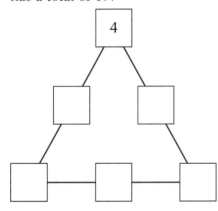

  b. Put the numbers 5, 6, 7, 8 and 9 in the boxes so that every row has the same total.

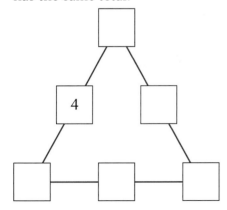

*27. $84 \times 36 = 3024$

Explain how you can use this fact to work out these calculations.

a. 84×72

b. 28×36

c. $3024 \div 18$

*28. The reciprocal of a whole number, N, is the fraction $\frac{1}{N}$.

For example, the reciprocal of 7 is $\frac{1}{7}$.

a. Write down the reciprocal of 9.

b. What number has a reciprocal of $\frac{1}{40}$?

c. Multiply the reciprocal of 3 by the reciprocal of 5.

d. Work out half of the reciprocal of 6.

*29. Textbooks cost £10.90 each.

There is a 10% discount if you buy more than 12.

Work out the cost of 24 books.

*30.

Chocolate bars

£1.69 each
3 for the price of 2

a. Work out the cost of six chocolate bars.

b. Sally bought some chocolate bars for her friends and she paid £11.83. How many did she buy?

*31. Here is the start of a sequence of fractions.

$$\frac{1}{3} \qquad \frac{2}{5} \qquad \frac{3}{7} \qquad \frac{4}{9} \qquad \frac{5}{11} \qquad \frac{6}{13} \qquad \cdots$$

a. Write down the next fraction in the sequence.

b. Write down the second fraction as a decimal.

c. Work out the sum of the first and second fractions. Write your answer as a fraction. Show your working.

*32. Here is the start of a sequence of patterns.

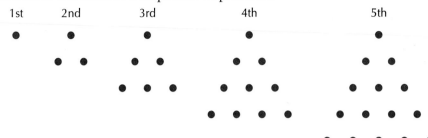

Serina says: 'The twentieth pattern has an even number of dots.'

Show that this is true.

*33. $\sqrt{A} = 4$

$\sqrt{B} = 3$

Rose says: 'In that case $\sqrt{A + B} = 7$.'

Show that Rose is wrong.

*34. This is a notice in a bookshop.

Buy **three** books and get the cheapest **free**

Books in the offer cost £4.99, £5.99 or £6.99.

Sam has £25 to spend.

a. Work out the largest number of books Sam can buy.

b. What is the largest amount of money Sam can save with the offer?

*35. Here is the start of a sequence of numbers.

5   17   11   14   …

To work out the next number in the sequence, add up the last two and divide by 2.

For example, $(5 + 17) \div 2 = 11$.

a. Work out the next three numbers in the sequence.

b. What can you say about the sequence as you calculate more numbers?

*36. Here is a list of numbers.

30   31   32   33   34   35   36   37   38

One set of three of these numbers that add up to 100 is 31 + 33 + 36.

Find all the other sets of three of these numbers that add up to 100.

2 Algebra

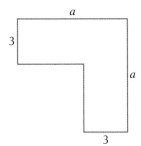

Show that the area of this shape is $6a - 9$.

SOLUTION

Divide the area into rectangles. Work out the separate areas and add them up.

There are several ways to do this. You can use whichever method you prefer.

Method 1

There are two rectangles here.

One area is $3 \times a = 3a$.

The other is $3 \times (a - 3) = 3(a - 3)$.

Add these and simplify the result.

$$3a + 3(a - 3) = 3a + 3a - 9 = 6a - 9$$

Because you have been given the answer you must show these steps.

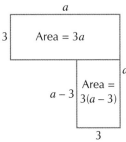

Method 2

There are three rectangles here.

The total area is $3(a - 3) + 9 + 3(a - 3) = 3a - 9 + 9 + 3a - 9 = 6a - 9$ as before.

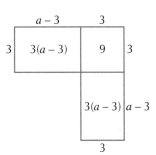

1. Raina is solving the equation $2(x - 3) = 8$.

This is her solution.

$2(x - 3) = 8$

$2x - 3 = 8$

$2x = 5$

$x = 2.5$

Raina has made some mistakes.

Write a corrected version of Raina's solution.

2. This question is about the expression $6x + 4y$.

a. Work out the value of $6x + 4y$ when $x = 2$ and $y = 9$.

b. Find a different pair of values of x and y which give the same value of $6x + 4y$.

3. These are the first four rectangles in a sequence.

   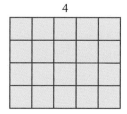

Brett says: 'The number of squares in the nth rectangle is $n(n + 1)$.'

a. Show that Brett's formula is correct when $n = 2$ and when $n = 4$.

b. Use Brett's formula to work out the number of squares in the 19th rectangle.

4. Look at this graph.

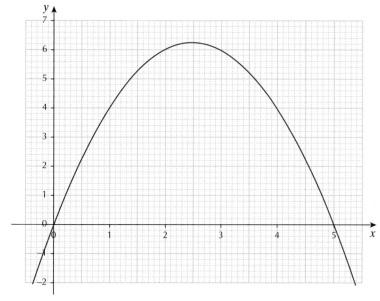

The equation of the graph is:

A. $y = x^2 - 5$ **B.** $y = 5 - x^2$ **C.** $y = x(x - 5)$
D. $y = x(5 - x)$ **E.** none of these.

Choose the correct answer.

Give a reason for your choice.

5. Square tables are put in a row and chairs are put round them.

This diagram shows the arrangement for three tables.

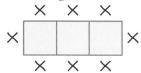

The chairs are represented with crosses.

a. How many chairs are needed for four tables?

b. Cameron says: 'If there are t tables you need $2(t + 1)$ chairs.'

Show that this formula is correct for the arrangement shown.

c. Use the formula to work out how many chairs eight tables need.

d. A particular arrangement uses 24 chairs. How many tables are there?

e. There are 35 chairs available all together.

Peta says: 'No arrangement can use all the tables with none left over.'

Show that Peta is correct.

6. Jason says: 'If x is a number then $2x + 1$ is always bigger than x.'

Show that this is false.

7. Look at this sequence of fractions.

$$\frac{1}{4} \qquad \frac{2}{5} \qquad \frac{1}{2} \qquad \frac{4}{7} \qquad \frac{5}{8} \qquad \frac{2}{3} \qquad \frac{7}{10}$$

Mel says: 'The sequence of fractions follows a pattern.'

Nick says: 'The fractions do not follow a pattern because $\frac{1}{2}$ and $\frac{2}{3}$ do not fit.'

Who is right?

Justify your answer.

8. Look at this flow chart.

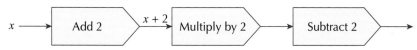

a. Write down an expression for the outcome of the flow chart.

b. Show that your expression is the same as $2(x + 1)$.

9. $x = 4$ and $y = 6$

Decide whether each statement below is true or false. Give a reason for your answer each time.

a. $(x + 1)(y + 1)$ is an odd number.

b. $(x + y)^2 = 52$

c. $(3x)^2 = (2y)^2$

10. a. Write down the next two terms of each of these sequences.

    i. 1 2 4 8 16 …

    ii. 3 5 7 9 11 …

b. Use the sequences in part **a** to find the next term in each of these sequences.

    i. 4 7 11 17 27

    ii. 5 9 15 25 43

    iii. 3 10 28 72 176

11.

The width of a rectangle is 8 cm and the length is y cm.

a. Aaron says: 'The area of the rectangle is 124 cm^2.'
Write an equation in y to express this.

b. Betina says: 'The perimeter of the rectangle is 47 cm.'
Write an equation in y to express this.

c. Work out the value of y.

12. This is the start of an arithmetic sequence.

    10   14   18   22   26   …

Is 100 in this sequence? Give a reason for your answer.

13. Gina has £t.

Danny has £10 less than Gina.

Martin has twice as much as Gina.

a. Write an expression for the total amount that all three have.

b. All together they have £200. Write an equation to show this.

c. Work out how much Danny has.

14. $a = 5$ and $b = 2$

    a. Work out the difference between a^2 and b^2.

    b. Work out the difference between $\frac{a}{b}$ and $\frac{b}{a}$.

15. a. These shapes have the same perimeter. Work out the value of x.

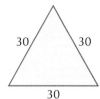

    b. These shapes have the same perimeter. Work out the value of y.

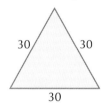 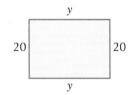

    c. These shapes have the same perimeter. Work out the value of z.

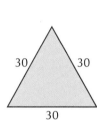 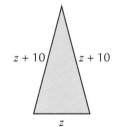

16. The nth term of one sequence is $2n + 10$.

    The nth term of another sequence is $3n - 2$.

    a. Show that the number 22 is in both sequences.

    b. Simon says: 'The number 22 is the only one that appears in both sequences.'

    Show that Simon is not correct.

17. In a Fibonacci sequence, each term (after the first two) is the sum of the previous two.

    a. These are the first five terms of a Fibonacci sequence.

        4   7   11   16   27

    Work out the seventh term.

    b. These are the first three terms of a Fibonacci sequence.

        x y $x + y$

    Show that the sixth term is $3x + 5y$.

18. Alex buys a plant and measures its height every week.

The height, y cm, after he has had the plant for x weeks, is given by the formula $y = 0.5x + 12$.

a. Work out the height of the plant after 4 weeks.

b. Work out the height of the plant after 8 weeks.

c. Draw a graph to show the increasing height of the plant. Use axes like this.

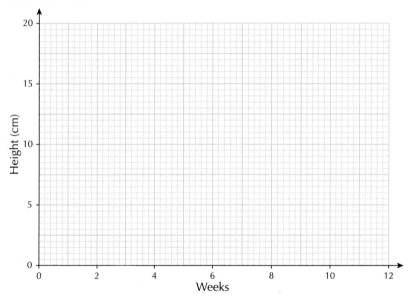

d. How does the graph show the height of the plant when Alex bought it?

e. Alex says:

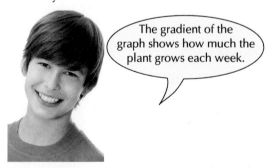

The gradient of the graph shows how much the plant grows each week.

Show that Alex is correct.

19. This is a rectangle.

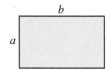

The width is *a* units and the length is *b* units.

a. Work out an expression for the perimeter.

b. Two rectangles identical to the one above are put together.

Work out an expression for the perimeter of this rectangle.

c. The two rectangles are put together in a different way.

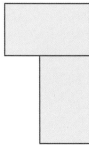

Work out an expression for the perimeter of this rectangle.

d. The two rectangles are put together to make this shape.

Work out an expression for the perimeter of this shape.

20. a. Match the words to the correct algebraic expression.

Think of a number, double it and add 5.

$x + 5$

Think of a number and add 10 to double the number.

$2x + 10$

Think of a number and double the answer after you have added 5.

$2(2x + 5)$

Think of a number and add five to it.

$2x + 5$

$2(x + 5)$

b. One algebraic expression does not have an explanation in words. Write one for it.

c. Two of the expressions will always give the same answer, whichever number you think of. Which are they?

21. Dave and Maggie choose a number together.

Dave says: 'I have multiplied our number by 4 and added 10.'

Maggie says: 'I have added 30 to our number.'

They both have the same answer.

  a. Use x to stand for their number and write down an equation it must satisfy.

  b. Work out the number they chose originally.

22. Look at these six formulae. Some of them are equivalent. Group the equivalent formulae together.

A: $x = c - 2 - y$ B: $c = y - x - 2$ C: $y + c = 2 - x$

D: $y = c - x - 2$ E: $x + y + c = 2$ F: $x + y = c - 2$

23. Match each sequence to the correct formula for its nth term.

A. 2 6 10 14 …		$3n$
B. 2 6 12 20 …		$2(n + 1)$
C. 3 6 11 18 …		$n^2 + n$
D. 3 6 9 12 …		$4n - 2$
E. 4 6 8 10 …		$n^2 + 2$

24. Alan says:

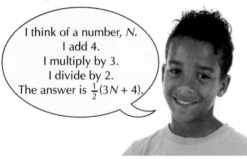

I think of a number, N.
I add 4.
I multiply by 3.
I divide by 2.
The answer is $\frac{1}{2}(3N + 4)$.

  a. Show that Alan's expression is not correct.

  b. Write a correct expression for Alan's answer.

*25. N is a positive whole number.

  a. Nina says: '$2N + 1$ is always a prime number.'

  Show that this is false.

  b. Carlos says: '$N^2 + N + 1$ is always a prime number.'

  Is this true or false?

  Give a reason for your answer.

*26. $(x + 1)^2 = x^2 + 2x + 1$

    a. Write each of these expressions in a similar way.

       i $(x + 2)^2$ **ii** $(x + 3)^2$

    b. a is a whole number and $(x + a)^2 = x^2 + 12x + 36$.

       Work out the value of a.

    c. c and d are whole numbers and $(x + c)^2 = x^2 + 18x + d$.

       Work out the values of c and d.

    d. e and f are whole numbers and $(x + e)^2 = x^2 + fx + 49$.

       Work out the values of e and f.

    e. Marcus says: 'I can find whole numbers n and m so that $(x + m)^2 = x^2 + nx + 75$.'

       Show that Marcus is wrong.

*27. Here are two algebraic expressions.

    $3y - 5$ $y + 2(y - 3)$

    a. Work out the value of each expression when $y = 12$.

    b. Gemma says: 'When you choose any number for y the values of the two expressions will differ by one.'

       Show that Gemma is correct.

*28. a. Simplify this expression. $2 + 5g - 3(g + 6)$

    b. Solve this equation. $2 + 5g - 3(g + 6) = 21$

    c. Solve this equation. $2 + 5g = 3(g + 6)$

*29. Here are the first four patterns in a sequence.

  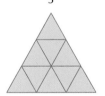 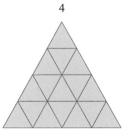

    a. Write down the number of small triangles in each pattern.

    b. How many small triangles are there in pattern 14?

    c. How many small triangles do you add to pattern 3 to make pattern 4?

    d. How many triangles do you add to pattern 14 to make pattern 15?

*30. The coordinates of the centre of a square are $(1, -2)$.

The coordinates of one corner of the square are $(4, 1)$.

Work out the coordinates of the other three corners.

*31. a.

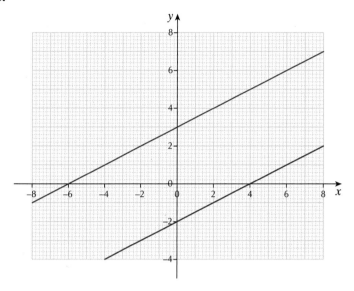

The equation of the blue line is $y = \frac{1}{2}x - 2$.

Work out the equation of the red line.

b.

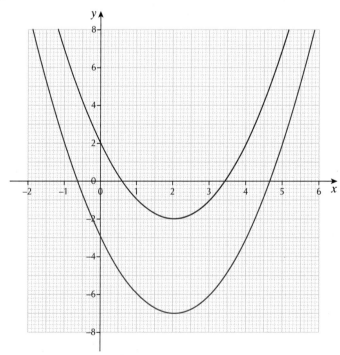

The equation of the blue line is $y = x^2 - 4x + 2$.

Work out the equation of the red line.

*32. a. Given that $2x - 1 = 21$ what can you say about the value of x?

b. Given that $2x - 1 > 21$ what can you say about the value of x?

*33. The term-to-term rule for a sequence is 'multiply by two and add four'.

The third term is 32.

a. Work out the fourth term.

b. Work out the first term.

*34. The value of $2a + b$ is 20.

The value of $a + 2b$ is 13.

a. Work out the value of $3a + 3b$.

b. Work out the value of $a + b$.

c. Work out the value of a.

*35.

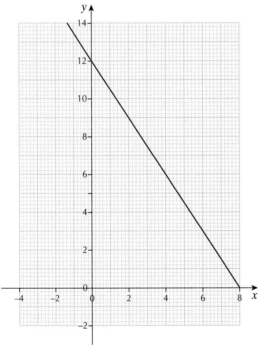

a. Show that the equation of this line is $3x + 2y = 24$.

b. Show that the equation can be written as $y = -1.5x + 12$.

3 Ratio, proportion and rates of change

This information is on the label of a can of chick peas.

Per 80 g serving	
Protein	6.2 g
Carbohydrate	13.2 g
Fat	1.1 g

a. Work out the ratio of carbohydrate to fat.

b. The can contains 250 g of chick peas.

Work out the amount of protein in a can.

SOLUTION

a. Use the masses to find the ratio.

The ratio of carbohydrate to fat is 13.2 : 1.1

Simplify this ratio. You can multiply both numbers by 10 to get whole numbers.

13.2 : 1.1 = 132 : 11

Both 132 and 11 are multiples of 11. Divide by 11.

The ratio is 12 : 1.

b. The amount of protein is proportional to the mass of chick peas. It is useful to put the numbers in a table.

Chick peas (g)	80	250
Protein (g)	6.2	

The multiplier for the chick peas is 250 ÷ 80 = 3.125.

The amount of protein in a can is 6.2 × 3.125 = 19.4 g.

The answer has been rounded to one decimal place.

1. The capacity of a small juice carton is 200 ml.

A pack of three cartons costs £0.87.

What is the cost per litre of the juice?

2. There are red and yellow tiles in this pattern.

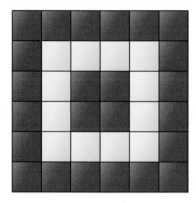

a. Work out the ratio of red tiles to yellow tiles. Write your answer as simply as possible.

b. What fraction of the tiles are red?

c. Write the number of yellow tiles as a percentage of the number of red tiles.

3. This is the label on a packet of beef.

Weight	Price per kg
0.274 kg	_____
PRICE	£4.31

Work out the missing price per kilogram.

4. The line AB is 20 cm long.

A   B

20 cm

C is on AB and AC : CB = 3 : 1.

D is on AB and AD : DB = 2 : 3.

Work out the length of CD.

5. 13% of my weekly pay is £34.19.

Work out 39% of my weekly pay.

6. A recipe uses these ingredients.

125 g Flour

50 g Sugar

25 g Butter

a. Write the ratio of the mass of sugar to the mass of flour, as simply as possible.

b. What fraction of the mass of flour is the mass of butter?

c. What percentage of the total mass of the three ingredients is butter?

d. The mass of flour is increased to 0.5 kg.

Work out the mass of sugar and butter needed to go with this.

7. Choose the units from m, cm, km and mm to make these statements correct.

a. 3.2 ____ = 320 ____

b. 0.76 ____ = 760 ____

c. 48 ____ = 4.8 ____

d. 0.9 ____ = 90 000 ____

8. A label on a pack of cheese shows this information.

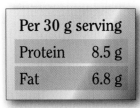

Per 30 g serving	
Protein	8.5 g
Fat	6.8 g

a. Work out what percentage of the mass of the cheese is fat.

b. Work out the ratio of protein to fat in a piece of cheese.

c. A sandwich contains 18 g of cheese. Work out the mass of fat in the cheese in this sandwich.

9. Rory draws some rectangles.

Top

Side

This table shows the top and side lengths for each one.

Top (cm)	8	9	10	16	18	24
Side (cm)	18	16	14.4	9	8	6

a. Draw axes like this and plot Rory's measurements. Join the points with a smooth curve.

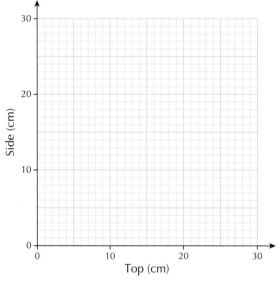

b. Show that the lengths of the tops and sides of Rory's rectangles are in inverse proportion.

c. Rory draws a square to go with his rectangles. Find the length of each side.

10. Look at this price list.

Pasta prices

| 250g bag | £0.95 |
| 450g bag | £1.53 |

Show that the larger bag is better value.

11. Carla lends her friend £200.

    Her friend pays Carla simple interest each month.

    At the end of six months Carla's friend has paid her £36 interest.

    Work out the monthly rate of interest.

12. a. Show that the ratio of $\frac{1}{2}$ to $\frac{3}{4}$ is 2 : 3.

    b. Show that the ratio of $\frac{1}{3}$ to $\frac{4}{9}$ is the same as the ratio of $\frac{1}{2}$ to $\frac{2}{3}$.

13. This information appears on the label on a one-litre bottle of squash.

Dilute to taste.
We suggest one
part squash to
four parts water.

Using the suggested amount, work out how many 250 ml drinks can be made from the bottle of squash.

14. The graph can be used to convert British pounds to Japanese yen.

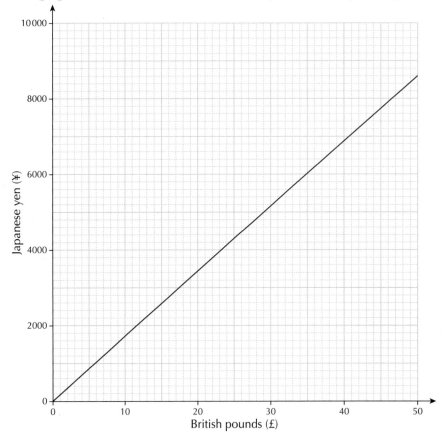

The exchange rate is the number of Japanese yen you can get for one pound.

Use the graph to work out the exchange rate.

15.

A million seconds is less than a fortnight

Is Beth right? Give a reason for your answer.

16. Jasmine and Kim have £68 between them.
Jasmine has $1\frac{1}{2}$ times as much as Kim.
Work out how much each person has.

17. Milk is sold in containers of different capacities.
The capacities are given in litres and pints.
Here are the labels from three different milk containers.

A   B   C

4 pints	**1** pint	___ pints
2.272 litres	_____ litres	1 litre

  a. Work out the missing number of litres on label B.

  b. Work out the missing number of pints on label C.

18. a. Copy this rectangle and shade $\frac{5}{8}$ of the squares.

  b. Work out the ratio of shaded squares to unshaded squares.

  c. How many more squares must you shade to make the ratio of shaded to unshaded 3 : 1?

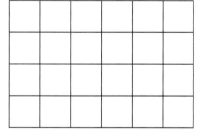

19. Saleema swims 20 lengths of a pool in 12 minutes.
The length of the pool is 25 metres.

  a. Work out Saleema's speed in metres/minute.

  b. How long will it take her to swim one kilometre at the same speed?

  c. Amir swims 20 lengths.
The ratio of Amir's time to Saleema's time is 4 : 3.
How much longer than Saleema does Amir take?

20. Rice is sold in bags in three different sizes.

Size	Small	Medium	Large
Mass	250 g	600 g	1.5 kg
Price	£0.65	£1.56	£3.70

a. Work out the ratio of the masses of a medium bag and a large bag.

b. What percentage of the mass of the medium bag is the mass of the small bag?

c. Is price proportional to mass? Give a reason for your answer.

21. A piece of tin has a mass of 19.9 g and its volume is 2.6 cm^3.

A piece of copper has a mass of 37.0 g and its volume is 4.1 cm^3.

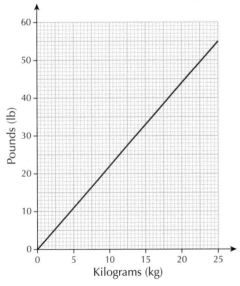

Which metal has the greater density? Give a reason for your answer.

22. Mass can be measured in kilograms (kg) or pounds (lb).

a. Explain how the graph shows that mass in pounds and mass in kilograms are proportional.

b. Mark has a mass of 70 kg. Work out his mass in pounds.

*23. George is using a rowing machine.

In six minutes the machine registers that he has rowed 930 metres.

a. If he rows at the same rate, how far will he row in ten minutes?

His target is to increase the distance he can row in six minutes by 10%.

b. What is his target, in metres?

The next time he uses the machine it registers that he has rowed 1075 metres in six minutes.

c. Work out the percentage increase.

*24. This graph shows the mass of A4 sheets of paper.

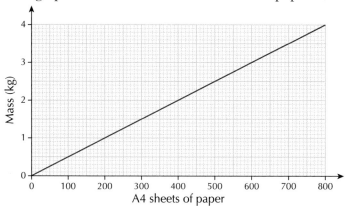

a. Explain how the graph shows that mass is proportional to the number of sheets.

b. A4 paper is sold in packs of 500. Find the total mass of four packs.

c. The mass of paper is given on the pack in grams per square metre (g/m²).

The area of an A4 sheet is $\frac{1}{16}$ m².

Work out the mass in g/m².

*25. A statement on a box of cereal says that a 30 g serving contains 6.9% of the guideline daily amount (GDA) of protein.

a. Work out the percentage of the GDA in a 50 g serving of cereal.

b. How much cereal gives the total GDA of protein?

*26. a. Anton can jog 100 m in 20 s. Work out his speed in metres per second (m/s).

b. This graph converts speeds in m/s to speeds in kilometres per hour (km/h).

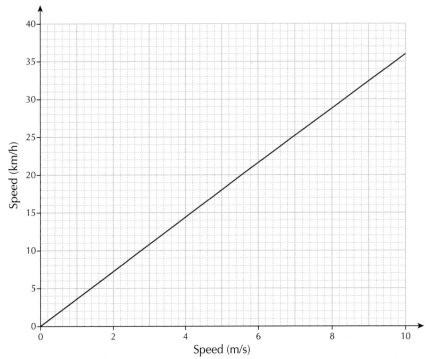

Use the graph to convert Anton's speed into km/h.

c. 1 mile ≈ 1.6 kilometres

Use this fact to convert Anton's speed into miles/hour.

*27. Gregor is growing a pumpkin. He weighs it every two weeks. Here are some of his results.

Date	1 August	15 August	29 August
Mass (kg)	5.0	5.5	6.0

Gregor makes three statements.

A: The increase in mass from 1 to 15 August is 10%.

B: The increase in mass from 15 to 29 August is 10%.

C: The increase in mass from 1 to 29 August is 20%.

Show that two of these statements are true and one is false.

*28. Alyson and Berta share £80 in the ratio 3 : 2.

Then Alyson gives Berta £4.

a. Work out the percentage of the money that Berta has.

b. Each of them spends £20. What fraction of Alyson's amount is Berta's amount now?

*29. Men and women take part in a half-marathon race.

65% of the runners are men.

  a. Work out the ratio of men to women.

  b. Work out the ratio of the number of women to the total number of runners.

  c. What fraction of the number of men is the number of women?

  d. There are 420 men in the race. Work out the total number of runners.

*30. This graph can be used to convert pounds into dollars or euros.

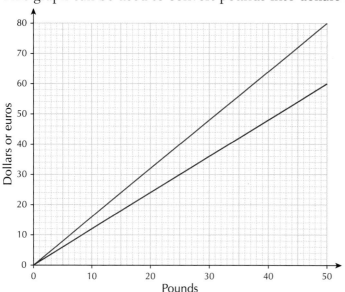

The red line shows dollars ($).

The blue line shows euros (€).

  a. The price of a watch is £50.

  Work out the price of the watch in dollars and in euros.

  b. The price of a handbag is $48. Work out the price in euros.

  c. Alison has $120. Does she have enough to buy a coat priced at €100? Give a reason for your answer.

*31. 40% of the people in a crowd are children.

60% of the children are girls.

What percentage of the crowd are girls?

*32. There are men, women and children in a park.

The ratio of men to women is 1 : 4.

The ratio of women to children is 1 : 3.

Work out the ratio of men to children.

*33. A cyclist is travelling at 21 km/hour.

How many metres does she travel in one minute?

*34. You can hire a boat on a river for a private party.

This graph shows the cost per person for different group sizes.

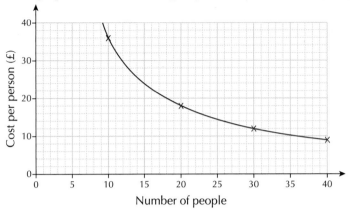

a. Explain how the graph shows that the cost per person is inversely proportional to the number of people.

b. Work out the cost per person for a group of 100 people.

*35. Electricity prices are being increased.

This graph shows electricity bills before and after the increase.

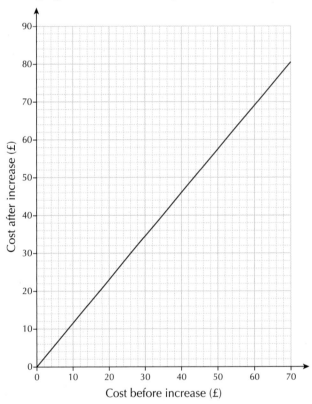

a. A bill for electricity before the increase is £60. What is the bill after the increase?

b. Work out the percentage increase in electricity bills.

c. The electricity bill for a family before the increase in price is £370. Work out the bill after the increase.

4 Geometry and measures

WORKED EXAMPLE

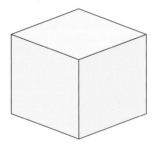

The area of each face of this cube is 64 cm^2.

Work out the volume of the cube.

SOLUTION

You should know how to find the area of a square and the volume of a cuboid.

You need to know the length of each side of the cube.

The area of a square face is 64 cm^2 so the length of each side is $\sqrt{64} = 8$ cm.

The volume is $8 \times 8 \times 8 = 512$ cm^3.

WORKED EXAMPLE

Here are two congruent equilateral triangles.

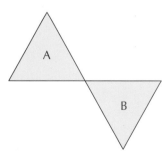

What transformations can map triangle A onto triangle B?

SOLUTION

The possible answers are a translation, a rotation or a reflection.

It cannot be a translation because the orientation of B is different to that of A.

It could be a rotation of 180° about the point where they touch.

(Remember to give the angle for a rotation.)

It could also be a reflection. The mirror line is the line of symmetry, shown in red.

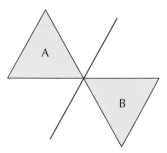

Always give the mirror line for a reflection.

---

1. The numbers on opposite faces of a dice add up to seven.

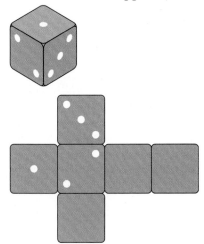

   Add the missing numbers to the net of the cube.

2. This is an isosceles triangle.

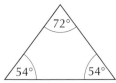

   a. Show how to put two of these triangles together to make a rhombus.

   b. Show how to put three of these triangles together to make a trapezium.

   c. Show how to put five of these triangles together to make a regular pentagon.

3. The line OP is rotated 90° clockwise about O and then reflected in the *y*-axis.

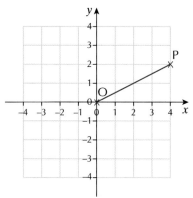

What are the final coordinates of point P?

4. This triangle is enlarged with a scale factor of 2.

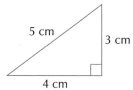

Phil says: 'The enlarged triangle is twice the size of the original triangle.'

Show that Phil's statement is not correct.

5. The diagram shows two squares drawn on a centimetre-square grid.

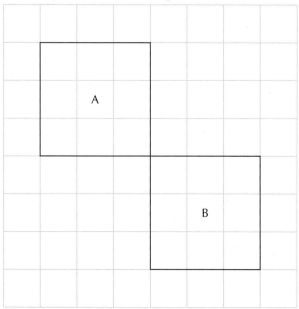

Describe three different transformations that map square A onto square B.

6. This diagram is drawn on a centimetre-square grid.

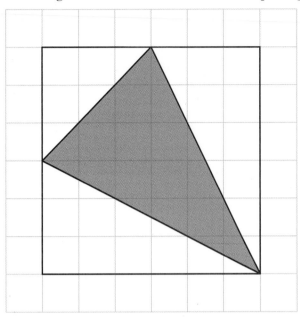

Work out the area of the red triangle. Show your working.

7. These are the names of four shapes.

cube   pyramid   triangular prism   cuboid

For each statement below, choose the possible name of the shape from the list. Give all possible answers – there may be more than one.

a. A shape with five faces

b. A shape with eight vertices

c. A shape with nine edges

8. Marta measures the angles marked $a°$ and $b°$ in the diagram.

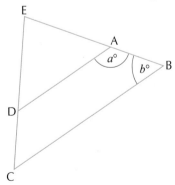

How can Marta use her measurements to decide whether lines AD and BC are parallel? Give a reason for your answer.

9. ABC is a right-angled triangle.

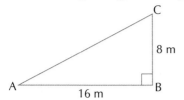

AB = 16 m and BC = 8 m

Show that 17 m < length of AC < 18 m.

10. This shape is drawn on a centimetre-square grid.

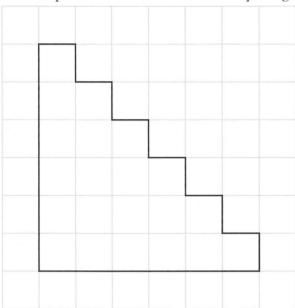

   a. Draw a rectangle with the same area as the shape.

   b. Draw a different rectangle with the same area.

   c. Draw a parallelogram with the same area.

   d. Draw a triangle with the same area.

   e. Draw a different triangle with the same area.

   f. Jason says: 'A circle with a radius of 2.6 cm has approximately the same area.'

    Is Jason correct? Justify your answer.

11. Here are a cuboid and a cube.

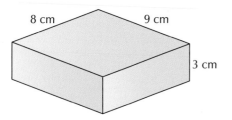 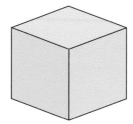

The cuboid and the cube have the same volume.

Work out the length of each side of the cube.

12. a. Jake finds the area of this shape by dividing it into two rectangles.

Show a way to do this. Work out the area.

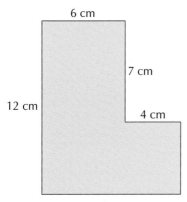

b. Farouk finds the area of the same shape by dividing it into two trapezia.

Show a way to do this. Show that Farouk's method gives the same answer.

13. a. Find a reflection followed by a translation that will map shape A onto shape B.

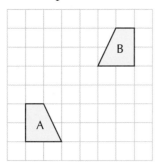

b. Is your answer to part **a** the only possible one? Give a reason for your answer.

14. This diagram is drawn on a centimetre-square dotted grid.

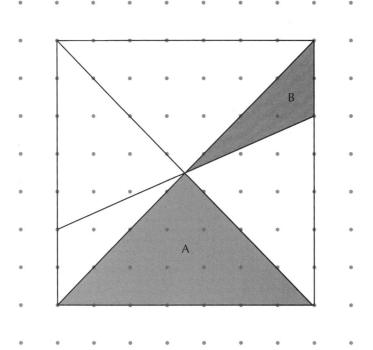

    a. Work out the area of triangle A.

    b. Work out the area of triangle B.

15. Here are some sets of three numbers.

    For each set, decide whether they could be the lengths of the sides of a right-angled triangle. Justify your answer each time.

    a. 8, 15, 17

    b. 8, 10, 6

    c. 8, 9, 12

16. Each side of a wooden cube is 3 cm long.

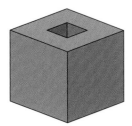
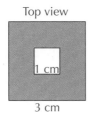

Top view

1 cm

3 cm

A square hole with a side of 1 cm is cut through the cube from the middle of one face to the middle of the opposite face.

Work out the volume of the remaining wooden shape.

*17. An equilateral triangle and a square both have a perimeter of 24 cm.

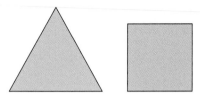

Show that the area of the triangle is less than the area of the square.

*18. Each angle of a regular polygon is 108°.

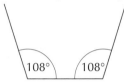

Show that the polygon is a pentagon.

*19. Work out the perimeter of this triangle.

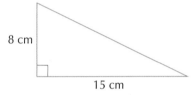

8 cm

15 cm

*20. Ollie draws some rectangles. All the measurements are in centimetres.

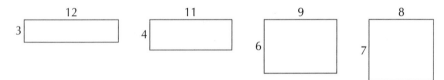

12

3

11

4

9

6

8

7

a. Show that all the rectangles have the same perimeter.

b. Show that the rectangles do not all have the same area.

c. Sketch a rectangle with the same perimeter but a smaller area than any of the original rectangles.

d. Ollie says: 'I can draw a square with the same perimeter.

What is the area of this square?

*21. The area of the top of a 2p coin is 5.3 cm^2.

Show that the diameter of the coin is 2.6 cm.

*22. a. One angle of a parallelogram is 75°.

Work out the sizes of the other three angles.

b. One angle of an isosceles triangle is 38°.

Work out all the possible sizes of the other two angles.

c. Two angles of a kite are 60° and 130°.

Work out all the possible sizes of the other two angles.

*23. These two congruent tiles are drawn on a centimetre-square dotted grid.

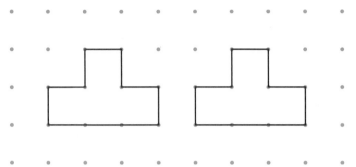

a. Here are two different ways to put them together with their edges meeting.

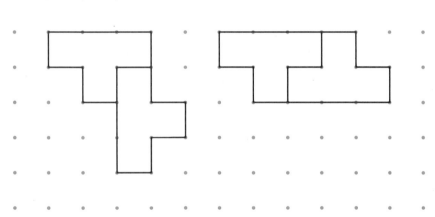

Show that the second shape has a smaller perimeter than the first.

b. You have four congruent tiles like the ones above.

Show how to put them together to give the smallest possible perimeter.

*24. The diameter of a bike wheel is 70 cm.

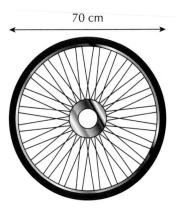

70 cm

How far does the bike move when the wheel goes round once?

*25.

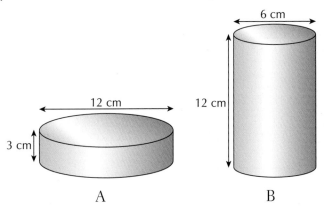

6 cm

12 cm

12 cm

3 cm

A

B

Celina says: Cylinder A has the larger volume.'

Jolanta says: 'Cylinder B has the larger volume.'

Who is correct? Justify your answer.

5 Probability

There are red, blue and black pens in a drawer.

Tracey takes one pen from the drawer, without looking.

The probability that she takes a red pen is $\frac{3}{5}$.

The probability that she takes a black pen is $\frac{3}{10}$.

a. Work out the probability that her pen is:

  i. not red   **ii.** blue.

b. Denise says: 'There are now 15 pens left in the drawer.'

  Explain why this is false.

SOLUTION

The question says that Tracey takes the pen without looking. This means that she is equally likely to take a pen of any colour. You could also say that she takes a pen 'at random'.

a. i. The probability that the pen is **not** red = 1 − the probability that it **is** red

$$= 1 - \frac{3}{5} = \frac{2}{5}$$

  ii. The pen must be red, blue or black. The three probabilities must add up to one.

  The probability the pen is blue $= 1 - \left(\frac{3}{5} + \frac{3}{10} \right)$

$$= 1 - \frac{9}{10} = \frac{1}{10}$$

b. If there are 15 pens left then originally there were 16 pens, since only one has been taken out.

  But originally $\frac{3}{5}$ of the pens were red — that is what the probability tells you.

  However, $\frac{3}{5}$ of 16 is not a whole number. Tracey must be wrong.

  You could also have used one of the other probabilities ($\frac{3}{10}$ or $\frac{1}{10}$) to justify your answer.

1. Graham has seven cards. Each card has a letter and a number on it.

| A 1 | B 2 | C 3 | D 4 | E 5 | F 6 | G 7 |

Graham takes a card at random.

Work out the probability that the card has on it:

a. a multiple of 3

b. a letter in the word GRAHAM

c. an even number and a letter in the word CAMERA.

2. Weather each day is put in one of three categories:

　　sunny

　　cloudy and dry

　　wet.

The probability it is sunny today is 0.3.

The probability it is not cloudy and dry is 0.9.

What is the probability it is wet?

3. Lucy has a large jar that contains 80 coloured sweets.

She says: 'If you take one without looking, the probability that you will **not** get a red sweet is $\frac{4}{5}$.'

Lucy is correct. How many of the sweets in the jar are red? Give a reason for your answer.

4. Gurdeep has a 2p coin, a 10p coin and a 20p coin.

He throws all three coins. Each coin can show a head (H) or a tail (T).

a. Copy and complete this table to show all the possible outcomes. You will need to add more rows.

2p	10p	10p
H	H	H

b. Work out the probability of his throwing at least one head.

c. Work out the probability of his throwing more heads than tails.

5. In a game, Sasha throws darts at this target.

The probability that Sasha will miss the target is 10%.

The probability he will hit red is twice the probability he will hit blue.

What is the probability he will hit red?

6. Ewan has a pack of cards.

Each card has on it a two-digit number.

Ewan takes a card at random.

The probability that it is an even number is $\frac{3}{4}$.

Look at the statements below. Say whether each one is true or false. Give a reason for each answer.

a. A quarter of the cards have odd numbers.

b. The number on Ewan's card cannot be 98.

c. There could be 50 cards in the pack.

7. This is part of a newspaper report.

> ## Lucky Seven
>
> Seven is people's favourite odd number.
>
> A scientist asked 250 people to choose an odd number less than ten. He found that 32% chose seven and 26% chose five.
>
> The other odd numbers were equally likely to be chosen.

What is the probability that a person chose the number three?

*8. There are 180 raffle tickets in five different colours.

This pie chart shows the proportion of each colour.

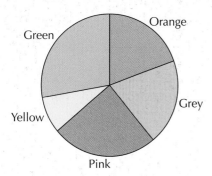

Leszek chooses one ticket at random.

a. Which colour is most likely?

b. Which two colours have the same probability?

c. The probability of one colour is $\frac{1}{12}$. Which colour is that?

d. What is the probability that the ticket is not pink?

*9. The probability that Hendrick is late for school is one-ninth of the probability that he is not late.

What is the probability that he is late for school?

*10. Cassie throws a dice twice and gets a 4 both times.

She is going to throw the dice again.

She says: 'The probability of a 4 this time is more than $\frac{1}{6}$.'

Is Cassie correct? Give a reason for your answer.

6 Statistics

WORKED EXAMPLE

This bar chart shows the ages of a group of people.

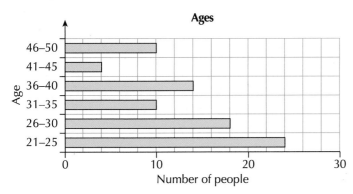

Choose the correct description, from the box below, for each statement.

| true | false | could be true or false |

a. There are 70 people in the group.

b. The median age is 28 years.

c. The range is 23 years.

SOLUTION

a. The sum of the frequencies gives the number of people in the group.

$24 + 18 + 10 + 14 + 4 + 10 = 80$

There are 80 people, so the statement is false.

b. There are 80 people.

The median age is half-way between the 40th and 41st ages when they are listed in order.

There are 24 people aged 25 or less.

There are $24 + 18 = 42$ people aged 30 or less.

This means that the 40th and 41st are both in the 26–30 group.

The median could be 28. The statement could be true or false.

c. The range is the difference between the ages of the oldest and youngest people.

The oldest is between 46 and 50.

The youngest is between 21 and 25.

The range could be as small as $46 - 25 = 21$ years.

The range could be as large as $50 - 21 = 29$ years.

23 is between 21 and 29 so the statement could be true or false.

1. This bar chart shows the results of a survey at an airport.

   It shows the destinations of some travellers.

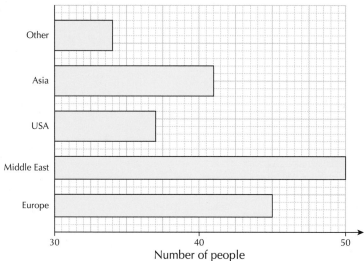

   There are errors in this bar chart.

   Draw a correct version.

2. Tickets are on sale for a concert. There are three different prices.

   This bar chart shows the numbers of tickets sold on three consecutive days.

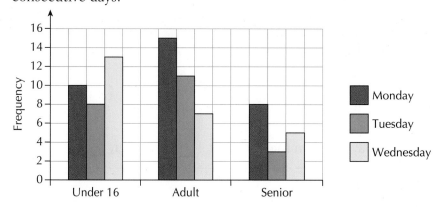

   Work out the total number of tickets sold on Monday.

3. This is the first line of a novel.

'IN A HOLE IN THE GROUND THERE LIVED A HOBBIT.'

This bar chart shows the frequency of each letter.

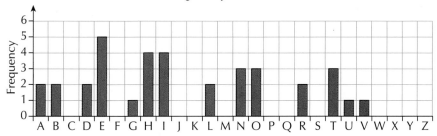

a. What is the modal letter?

b. Why is it not possible to work out the mean?

c. If the frequencies are shown in a pie chart, what is the angle for the letter E, to the nearest degree?

d. Give one reason why a pie chart is not a good choice to show the frequencies of letters.

4. This bar chart shows the numbers of boys and girls who go swimming after school each day.

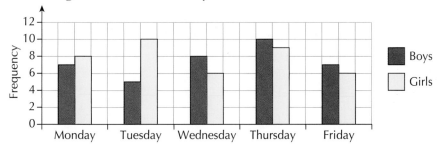

a. What fraction of the children who go on Tuesday are boys?

b. Show that the median number of boys is less than the median number of girls.

c. This pie chart shows the data for either boys or girls.

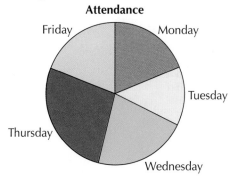

Is the pie chart for boys or for girls? Give a reason for your answer.

5. This table shows the number of goals scored in the 48 games of the first round of the 2014 football world cup.

Goals scored in the match	Number of matches
0	5
1	8
2	4
3	15
4	9
5	4
6	2
7	1

    a. In what fraction of the 48 games were more than three goals scored? Write your answer as simply as possible.

    b. Show that, in total, over 100 goals were scored.

6. A café sells coffee in three sizes: small, medium and large.

This bar chart shows the sales in one week.

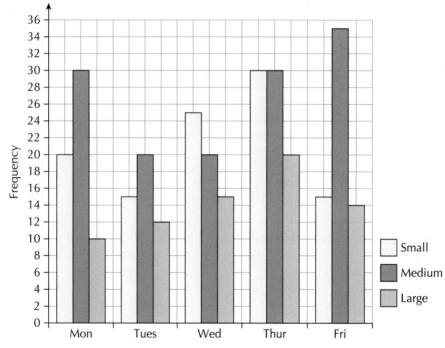

    a. Which size has the largest range of sales?

b. This pie chart shows the sales of one size of coffee.

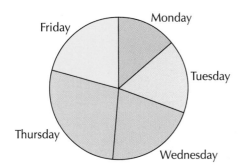

Which size is it? Give a reason for your answer.

c. This pie chart shows the totals for each size.

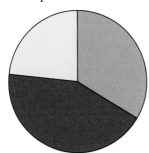

Which size is represented by the red sector?

7. The pie charts show the age distributions of residents in two towns.

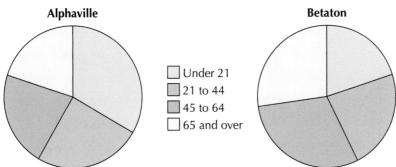

Say whether each statement is true, false or you cannot tell.

a. There were more under 21s in Alphaville.

b. The modal age group for both towns was the same.

c. The median age for both towns was the same.

d. The populations of both towns was the same.

e. There was a smaller proportion of people aged 45 and over in Alphaville than is Betaton.

*8. a. A group of 15 children each throw a dice once.

The mean of all their scores is 3.6.

Show that the total of all the throws is 54.

b. A second group of children each throw a dice once.

The mean score is exactly 3.1.

Carmen says: 'There cannot be 15 children in this group.'

Show that this is true.

*9. This bar chart shows the ages of 60 people on a college course.

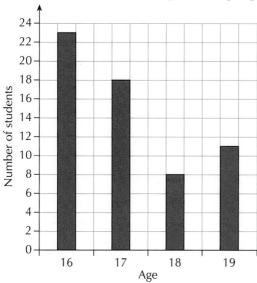

Carla says: 'A pie chart is a better way to show the proportion that are 16 years old.'

a. Explain why Carla is correct.

b. Work out the angle of the largest sector of the pie chart.

*10. The mean number of goals scored in five football matches is 3.

After another match the mean increases to 3.5.

How many goals were scored in the sixth match?

7 Mixed questions

This pie chart shows the number of patients attending a clinic in one week.

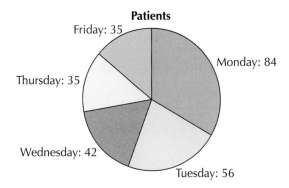

Patients

Friday: 35
Monday: 84
Thursday: 35
Wednesday: 42
Tuesday: 56

a. Work out the ratio of the numbers on Monday and Tuesday. Give your answer as simply as possible.

b. What percentage of the patients come to the clinic on Tuesday?

c. Work out the angle for the Friday sector.

SOLUTION

a. The ratio is 84 : 56.

Simplify the ratio by dividing by any common factor.

2, 7 and 14 are all common factors.

The simplest form is 3 : 2.

b. To work out the percentage you need to know the total for the week.

Total = 84 + 56 + 42 + 35 + 35 = 252

The percentage on Tuesday is $\frac{56}{252} \times 100 = 22.2\%$.

Round the calculator answer to 1 decimal place.

c. The total angle for the circle is 360°.

The angle on Friday is $\frac{35}{252} \times 360° = 50°$.

An alternative method is to say that the angle of 360° represents 252 patients.

1 patient is represented by 360° ÷ 252 = 1.428...°.

Do not round the calculator answer to this first calculation.

35 patients = 35 × 1.428...° = 50°

1. This table shows the prices of tickets for a football match.

	East stand	West stand	North stand	South stand
Adult	£28	£32	£23	£23
Under-16	£22	£25	£16	£16

There are four different stands for which you can buy tickets.

Mr Patel wants two adult and three under-16 tickets for his family.

Work out the difference in price between the most expensive and the least expensive stand.

2. Jasmine wants to draw a triangle. There are two rules.

The length of each side must be a whole number of centimetres.

The perimeter must be 24 cm.

Show how Jasmine can draw:

a. an equilateral triangle

b. an isosceles triangle

c. a scalene triangle.

3. Amber has a large number of tiles like this.

Show that Amber **cannot** put her tiles together to make this pattern.

4. This table shows the distances, in kilometres, between four European cities.

a. Which two cities are closest together?

b. Write the distance from Amsterdam to Brussels as a percentage of the distance from Amsterdam to Berlin.

c. A distance of 8 kilometres is approximately equal to 5 miles. Work out the approximate distance from Brussels to Amsterdam, in miles.

	Amsterdam	Berlin	Brussels	Paris
Amsterdam	✕	656	210	502
Berlin	656	✕	774	1053
Brussels	210	774	✕	311
Paris	502	1053	311	✕

5. There is a large number of grapes in a bowl. The grapes are either black or green.

The ratio of black grapes to green grapes is 3 : 5.

a. Meryl takes a grape at random. What is the probability that the grape is green?

b. Explain why there could be 40 grapes in the bowl but there could not be 50 grapes in the bowl.

c. After some grapes have been eaten, the ratio of black to green is 1 : 4.

What percentage of the remaining grapes are black?

6. This is a rectangular wall.

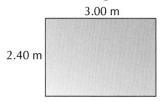

Josie is covering the wall with tiles.

The tiles are 15 cm square.

How many tiles does Josie need?

7. This chart shows the ages at which a group of men and a group of women were married.

a. What percentage of the women were aged 31-40?

b. What was the modal age class for men?

c. Gary draws a pie chart to show the ages of the men. What is the angle of the sector for men over 40 years old?

d. One of the women is chosen at random. What is the probability that she was 21 or over when she married?

e. There were 7500 men in the group. How many were over 30 when they married?

8. Cooks use teaspoons and tablespoons as measures.

> 1 teaspoon = 5 ml
> 1 tablespoon = 20 ml

a. How many tablespoons of milk are there in a one-litre carton?

b. The dose for cough mixture is three teaspoons.

How many doses are there in a 250 ml bottle?

c. What percentage of a tablespoon is three teaspoons?

d. Write the ratio of 2 teaspoons to 3 tablespoons as simply as possible.

9. Colin enlarges a rectangle by a scale factor of 3.

This is the size of the enlargement.

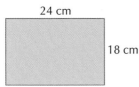

24 cm

18 cm

Work out the area of the original rectangle.

10. This question is about the whole numbers from 1 to 30 inclusive.

a. What fraction of the numbers are prime numbers?

b. Work out the ratio of the multiples of three to the multiples of five.

c. Work out the percentage of the numbers that are multiples of six.

d. Work out the percentage of the numbers that are factors of 30.

11. If N is a number, $\frac{1}{N}$ is called the reciprocal of N.

For example, the reciprocal of 7 is $\frac{1}{7}$.

a. The reciprocal of a number is 0.05.

What is the number?

b. Add together the reciprocal of 3 and the reciprocal of 4.

c. Work out 5 divided by the reciprocal of 6.

12. This table shows the result of a survey of whether men and women wear glasses.

	Wear glasses	**Do not wear glasses**
Men	8	22
Women	12	30

a. Work out how many people took part in the survey.

b. What percentage of the men in the survey wear glasses?

c. Among those who wear glasses, what is the ratio of men to women? Write your answer as simply as possible.

13. A fence is made from vertical posts and horizontal bars.

The posts are 3 metres apart.

This fence is 9 metres long. It has 4 posts and 6 bars.

a. Work out the number of posts and bars in a fence 15 m long.

b. Work out the length of a fence with 10 posts.

c. Is the number of posts proportional to the length of the fence? Give a reason for your answer.

d. Is the number of bars proportional to the length of the fence? Give a reason for your answer.

14. a. Find the sum of the factors of 30.

b. Show that 30 is the product of three prime numbers.

c. Write 30 as the sum of three prime numbers.

15. The graph shows the lines $y = 0.7x + 3.2$ and $y = 6.4 - 1.2x$.

Use the graph to find approximate solutions to these equations.

a. $0.7x + 3.2 = 6.5$

b. $6.4 - 1.2x = 2.3$

c. $0.7x + 3.2 = 6.4 - 1.2x$

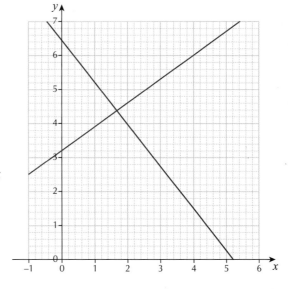

16. A car is travelling at 25 metres per second (m/s).

    a. How long does it take the car to travel 10 m?

    b. How long does it take the car to travel 10 km?

    c. The speed increases by 25%. Work out the new speed.

17. Jocasta is trying to solve this equation.

$2x + 5 = 5(x - 4)$

Here is Jocasta's attempt at a solution. She has made some mistakes.

$$2x + 5 = 5(x - 4)$$
$$2x + 5 = 5x - 9$$
$$3x + 5 = -9$$
$$3x = -4$$
$$x = -1\tfrac{1}{3}$$

Write out a corrected version of Jocasta's solution.

*18. a. Work out an expression for the perimeter of this shape.

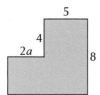

    b. Work out an expression for the area of the shape.

*19. a. The triangle in this diagram is equilateral.

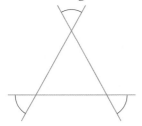

    Explain why each of the marked angles is 60°.

    b. The triangle in this diagram is not equilateral.

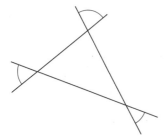

    Explain why the sum of the marked angles is 180°.

*20. Here are five numbers. The last two are the same.

    3   5   8 x x

    a. The mean of these five numbers is 9.

       Explain why $2x + 16 = 45$.

    b. Work out the two missing numbers.

*21. Shona uses Pythagoras' theorem to find the value of x in this triangle.

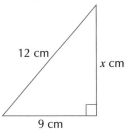

    This is what she writes.

    $x^2 = 12^2 + 9^2$

        $= 24 + 18$

        $= 42$

    $x = \sqrt{42} = 6.5$

    a. What mistakes has Shona made?

    b. Write out a correct version of her calculation.

*22. x and y are two numbers. The *geometric mean* of x and y is $\sqrt{xy}$.

    a. Show that the geometric mean of 2 and 8 is 4.

    b. Work out the geometric mean of 8 and 50.

    c. Two numbers have a geometric mean of 9.

       One of the numbers is 3.

       Work out the other number.

    d. Find all the pairs of positive whole numbers that have a geometric mean of 10.

*23. a. Show that the perimeter of this rectangle is 24 cm.

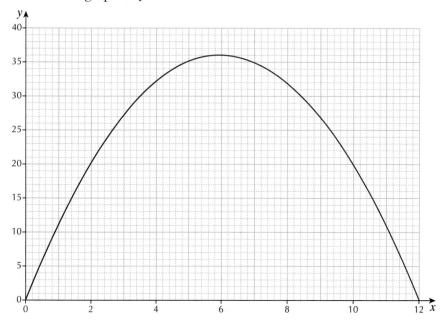

12 − x cm

x cm

b. Show that an expression for the area of the rectangle is $12x - x^2$ cm^2.

c. This is a graph of $y = 12x - x^2$.

i. Use the graph to find a value of x that will make the area approximately 30 cm^2.

ii. Explain why it is impossible for the area to be 40 cm^2.

*24. The area of triangle ABC = 143 cm^2.

AD = 16 cm   DB = 10 cm

Work out the length of CD.

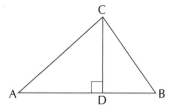

*25. The fourth term of an arithmetic sequence is 68.

The sixth term of the sequence is 80.

Work out the first term.

*26. Each side of a square is 10 cm long.

A circle touches each side of the square.

The corners are cut out and arranged to make this shape.

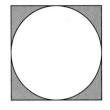

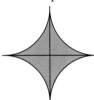

a. Work out the perimeter of the shape.

b. Work out the area of the shape.

*27.

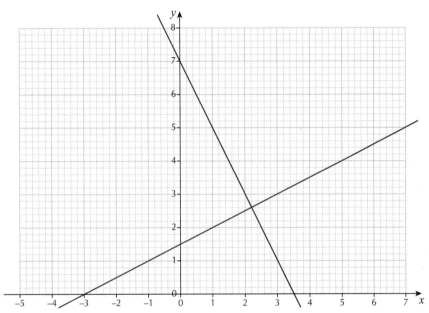

a. Show that the equations of these lines can be written as $2y = x + 3$ and $2x + y = 7$.

State which is line A and which is line B.

b. Work out the gradient of each line.

c. Use the graph to find an approximate solution to the simultaneous equations $2x + y = 7$ and $2y - x = 3$.

Notes

Solutions

The answers given include notes, where these are likely to be helpful.

1 Number

Qn	Answer	Notes
1	**a** 50p + 10p + 5p + 2p + 1p	
	b 40p	
2	**a** $-20.4°C$	
	b An increase of 2.2 degrees	
3	**a** 15, 51, 57 or 75	The two digits together must form a number that is a multiple of three. You can have 1 and 5 or 5 and 7 so the possible answers are 15, 51, 57 or 75.
	b 62	The largest is 75 and the smallest is 13. The difference is $75 - 13 = 62$.
4	Radio 4	The scale goes up in 2s. The arrow points to 93 so that is Radio 4.
5	**a** 45p	$£1.80 \div 4 = 45p$
	b 65p	$£5.05 - £1.80 = £3.25$ $£3.25 \div 5 = 65p$
6	**a** $\dfrac{1}{2}$	$\dfrac{3}{5} - \dfrac{1}{10} = \dfrac{6}{10} - \dfrac{1}{10} = \dfrac{5}{10} = \dfrac{1}{2}$
	b $\dfrac{7}{10}$	$\dfrac{1}{5} + \dfrac{1}{2} = \dfrac{2}{10} + \dfrac{5}{10} = \dfrac{7}{10}$
7	**a** Any multiple of 30.	30 is the least common multiple of 5 and 6.
	b Possible answers are 3, 6, 15, 30.	The factors of 30, in order, are 1, 2, 3, and you can stop there because 3 is not a factor of 50. Possible answers are 3, 6, 15, 30.
	c 56	$1 + 2 + 4 + 7 + 14 + 28 = 56$. Do not forget to include 28.
	d 100	Product means multiply. $1 \times 2 \times 5 \times 10 = 100$
8	0.3	0.3 because $6.1 - 2.3 = 3.8$ and $3.8 - 3.5 = 0.3$
9	**a** $5 + 4 \times (8 - 3)$	Remember to work out the brackets first. Try putting brackets in different places. $5 + 4 \times (8 - 3) = 5 + 20 = 25$
	b $(5 + 4) \times (8 - 3)$	$(5 + 4) \times (8 - 3) = 9 \times 5 = 45$
	c $(5 + 4) \times 8 - 3$	$(5 + 4) \times 8 - 3 = 72 - 3 = 69$
10	$34 = 25 + 9$; $37 = 36 + 1$; $40 = 36 + 4$; $41 = 25 + 16$; the one that cannot be done is 39.	The square numbers are 1, 4, 9, 16, 25, 36, … Showing how to make the other four is a good way to justify your answer.
11	**a** $\dfrac{3}{10} = \dfrac{6}{20}$ and $\dfrac{1}{4} = \dfrac{5}{20}$; $\dfrac{3}{10} = \dfrac{9}{30}$ and $\dfrac{1}{3} = \dfrac{10}{30}$	You can use equivalent fractions for this question.
	b One possible answer is $\dfrac{17}{24}$. There are many other possible answers e.g. $\dfrac{5}{7}$	$\dfrac{2}{3} = \dfrac{8}{12} = \dfrac{16}{24}$; $\dfrac{3}{4} = \dfrac{9}{12} = \dfrac{18}{24}$
12	**a** $-6, -4, -2, 3, 5$	If you are not sure, use a number line.
	b 11	Use the largest and smallest numbers. $5 - -6 = 11$

	c −2	−2 is the sum of the other four.
	d 24	24 = −6 × −4 Remember that the product of two negative numbers is positive.
13	**a** £44 900	£212 400 − £167 500 = £44 900
	b 25% is 0.25 × 167 500 = 41 875, less than 44 900 The increase is $\frac{44\ 900}{167\ 500} \times 100 = 26.8\%$, larger than 25%.	You can either find 25% of £167 500 or find the percentage increase.
14	11 and 13	This means $A \times B = 1001 \div 7 = 143$. Use a calculator to divide 143 by different prime numbers until you get a prime number answer. A and B are 11 and 13.
15	**a** 49 **b** 39 **c** 35 and 39	You should be able to recognise square numbers. It is 3 × 13 31, 37 and 41 are prime so it cannot be those. 5 × 7 = 35 and 3 × 13 = 39 so there are two answers.
16	£39	The calculation is 149 + (10 × 69) − 800 = £39.
17	**a** $\frac{1}{4}$ of 0.6 = 0.15; $\frac{3}{4}$ − 0.6 = 0.15 **b** $\frac{1}{5}$ of 0.5 = 0.1; $\frac{3}{5}$ − 0.5 = 0.1	$\frac{1}{4}$ of 0.6 = 0.6 ÷ 4 = 0.15; $\frac{3}{4}$ − 0.6 = 0.75 − 0.6 = 0.15 They are the same. $\frac{1}{5}$ of 0.5 = 0.5 ÷ 5 = 0.1; $\frac{3}{5}$ − 0.5 = 0.6 − 0.5 = 0.1 They are the same.
18	**a** 8 short sticks **b** 3 long sticks **c** Three short and two long or one short, one medium and two long	The rods are 250 cm. 250 ÷ 30 = 8 r 10 so 8 short sticks. 250 ÷ 80 = 3 r 10 so 3 long sticks. Try different arrangements. Three short and two long works. So does one short, one medium and two long.
19	**a** 5 + 23 or 11 + 17 **b** 2 + 3 + 23 or 2 + 7 + 19 **c** 3 + 5 + 7 + 13 **d** Yes. The five smallest prime numbers are 2 + 3 + 5 + 7 + 11 = 28.	28 = 5 + 23 or 11 + 17 One of them must be 2 in this case. 28 = 2 + 3 + 23 or 2 + 7 + 19. 28 = 3 + 5 + 7 + 13 is the only possible answer.
20	**a** 100 **b** 0.74 **c** 4600	The digits do not change, just the position of the decimal point. 4.6 × 10 = 46 so the missing number is 4600.
21	**a** $2\frac{1}{2}$ **b** 13 doses	$75 \div 30 = 2\frac{1}{2}$ 1000 ÷ 75 = 13 r 25 so the answer is 13 doses.
22	4 bottles of water	You need a combination of 90s and 260s to make 1140. The answer is 4 bottles of water.
23	£21.54 to the nearest penny.	The total cost is £35 × 8 = £280. Each pays £280 ÷ 13 = £21.54 to the nearest penny.
24	£697.50	15% = 0.15 × 1350 = £202.50 One-third = 1350 ÷ 3 = £450. The amount left is 1350 − 202.50 − 450 = £697.50

25	**a** 88% **b** £250	The amounts are halved in the second part. The amounts are trebled in the second part.
26	**a** **b**	The other numbers in the line with 4 must add up to 15. The other numbers in the line with 4 must add up to 16.
27	**a** $84 \times 72 = 84 \times 36 \times 2 = 3024 \times 2 = 6048$ **b** $28 \times 36 = 84 \times 36 \div 3 = 3024 \div 3 = 1008$ **c** $3024 \div 18 = 3024 \div 36 \times 2 = 84 \times 2 = 168$	
28	**a** $\dfrac{1}{9}$ **b** 40 **c** $\dfrac{1}{15}$ **d** $\dfrac{1}{12}$	 $\dfrac{1}{3} \times \dfrac{1}{5} = \dfrac{1}{15}$ $\dfrac{1}{6} \div 2 = \dfrac{1}{12}$
29	£235.44	The cost is 90% of $24 \times £10.90 = 0.9 \times £261.60 = £235.44$ An alternative method is to work out 10% and subtract it from the total.
30	**a** £6.76 **b** 10	6 cost $4 \times £1.69 = £6.76$ $11.83 \div 1.69 = 7$ 2×1.69 buys 3 bars so she bought $3 + 3 + 3 + 1 = 10$
31	**a** $\dfrac{7}{15}$ **b** 0.4 **c** $\dfrac{1}{3} + \dfrac{2}{5} = \dfrac{5}{15} + \dfrac{6}{15} = \dfrac{11}{15}$	
32	The numbers of dots are 1, 3, 6, 10, 15, 21, 28, … They are odd, odd, even, even, odd, odd, … and this pattern continues. So the 4th, 8th, 12th, 16th and 20th, are even.	Other explanations are possible.
33	$\sqrt{A} = 4$ means that $A = 4^2 = 16$. $\sqrt{B} = 3$ means that $B = 3^2 = 9$. $A + B = 25$ so $\sqrt{A + B} = 5$	
34	**a** 7 **b** £11.98	7 at £4.99 – two will be free £11.98 if he buys three at £6.99 and three at £4.99

35	**a** 12.5, 13.25 and 12.875	$25 \div 2 = 12.5$, $26.5 \div 2 = 13.25$ and $25.75 \div 2 = 12.875$
	b The numbers go up and down in value and get closer and closer to 13.	
36	$30 + 32 + 38$ $31 + 32 + 37$ $32 + 33 + 35$ $30 + 33 + 37$ $31 + 33 + 36$ $30 + 34 + 36$ $31 + 34 + 35$	List them systematically to make sure you find them all. There are seven sets.

2 Algebra

1	$2(x - 3) = 8$ $2x - 6 = 8$ multiply both terms by 2 $2x = 14$ add 6 to both sides $x = 7$ divide by 2	
2	**a** 48	$12 + 36 = 48$
	b There are many different values. Some are shown in this table.	

x	0	1	3	4	6	8	10
y	12	10.5	7.5	6	3	0	−3

3	**a** When $n = 2$, $n(n + 1) = 6$; when $n = 4$, $n(n + 1) = 20$. These are correct.	
	b 380	$19 \times 20 = 380$
4	D This is the only one to give these correct points: $(0, 0)$, $(5, 0)$, $(1, 4)$	
5	**a** 10	
	b When $t = 3$, $2(t + 1) = 2 \times 4 = 8$ which is correct.	
	c 18	$2(8 + 1) = 18$
	d 11	$2(t + 1) = 24$ so $t = 11$
	e When $2(t + 1) = 35$ then $t = 16.5$ You cannot have half a table.	
6	It is true for positive values of x but not for most negative ones. For example, when $x = -3$ then $2x + 1 = -5$ which is less than x.	
7	Mel is right if you write $\frac{1}{2}$ and $\frac{2}{3}$ as the equivalent fractions $\frac{3}{6}$ and $\frac{6}{9}$.	
8	**a** $2(x + 2) - 2$	
	b $2(x + 2) - 2 = 2x + 4 - 2 = 2x + 2 = 2(x + 1)$	
9	**a** True. $(x + 1)(y + 1) = 5 \times 7 = 35$	
	b False. $(x + y)^2 = 10^2 = 100$	
	c True. $(3x)^2 = 12^2 = 144$ and $(2y)^2 = 12^2 = 144$	
10	**a i** 32, 64	Double each time.
	ii 13, 15	Odd numbers
	b i 45	$32 + 13 = 45$
	ii 77	$2 \times 32 + 13 = 77$
	iii 416	$32 \times 13 = 416$

11	**a** $8y = 124$	
	b $2(8 + y) = 47$	Or an equivalent equation such as $2y + 16 = 47$
	c The solution of either equation is $y = 15.5$	
12	No. Here is one explanation. The nth term is $4n + 6$. If the nth term is 100 then $4n + 6 = 100 \rightarrow n = 23.5$ which is not a whole number.	You could also say that all the terms have a remainder of 2 when divided by 4 and 100 has no remainder, so 100 cannot be in the sequence.
13	**a** $t + (t - 10) + 2t$	You could simplify this to $4t - 10$.
	b $t + (t - 10) + 2t = 200$ or $4t - 10 = 200$	
	c $t = 52.5$ so Danny has £42.50	
14	**a** 21	$25 - 4 = 21$
	b 2.1	$\dfrac{5}{2} - \dfrac{2}{5} = 2.5 - 0.4 = 2.1$
15	**a** 22.5	$4x = 90 \rightarrow x = 90 \div 4 = 22.5$
	b 25	$2y + 40 = 90 \rightarrow 2y = 50 \rightarrow y = 25$
	c $23\dfrac{1}{3}$	$2(z + 10) + z = 90 \rightarrow 3z + 20 = 90 \rightarrow$ $3z = 70 \rightarrow z = 23\dfrac{1}{3}$
16	**a** When $n = 6$, $2n + 10 = 22$; when $n = 8$, $3n - 2 = 22$	
	b The first starts 12, 14, 16, 18, ... and the second starts 1, 4, 7, 10, 13, 16, ... so 16 is in both sequences. In fact the numbers 16, 22, 28, 34, ... are in both.	
17	**a** 70	The 6th is $16 + 27 = 43$; the 7th is $27 + 43 = 70$.
	b The 4th is $y + (x + y) = x + 2y$; the 5th is $(x + y) + (x + 2y) = 2x + 3y$; the 6th is $(x + 2y) + (2x + 3y) = 3x + 5y$	
18	**a** 14 cm	$0.5 \times 4 + 12 = 14$ cm
	b 16 cm	$0.5 \times 8 + 12 = 16$ cm
	c	
	d At 0 weeks the height is 12 cm	
	e The gradient is the coefficient of x, 0.5, and the height increases by 0.5 cm each week.	
19	**a** $2(a + b)$ or $2a + 2b$	
	b $2(a + 2b)$ or $2a + 4b$	
	c $2(2a + b)$ or $4a + 2b$	
	d $2a + 4b$	Or an equivalent expression
20	**a** 'Think of a number, double it and add 5' $\rightarrow 2x + 5$	
	'Think of a number and add 10 to double the number' $\rightarrow 2x + 10$	
	'Think of a number and double the answer after you have added 5' $\rightarrow 2(x + 5)$	
	'Think of a number and add five to it' $\rightarrow x + 5$	

	b $2(2x + 5) \rightarrow$ 'Double the number, add 5, then double the answer.' **c** $2x + 10$ and $2(x + 5)$	
21	**a** $4x + 10 = x + 30$ **b** $3x = 20 \rightarrow x = 6\frac{2}{3}$	
22	A, D and F are equivalent; C and E are equivalent	
23	A is $4n - 2$; B is $n^2 + n$; C is $n^2 + 2$; D is $3n$; E is $2(n + 1)$	Start with the easy formulas such as $3n$ and find terms.
24	**a** Show that the formula gives the wrong answer. For example, if $N = 1$ the answer is 7.5 but the formula gives 3.5. **b** $\dfrac{3(N + 4)}{2}$ or an equivalent expression such as $\dfrac{3}{2}(N + 4)$.	
25	**a** When $N = 4$, $2N + 1 = 9$ which is not a prime number. The answer is not always a prime. Any value that does not give a prime number is possible. **b** False. When $N = 4$, $N^2 + N + 1 = 21$ which is not a prime number. Other values are possible too.	
26	**a i** $(x + 2)^2 = x^2 + 4x + 4$ **ii** $(x + 3)^2 = x^2 + 6x + 9$ **b** 6 **c** $c = 9$ and $d = 81$ **d** $e = 7$ and $f = 14$ **e** You cannot find a whole number m so that $m^2 = 75$.	a^2 is 36 so $a = 6$ $2c = 18$ so $c = 9$ and $d = 9^2 = 81$
27	**a** 31 and 30 **b** $y + 2(y - 3) = y + 2y - 6 = 3y - 6$. This is one less than $3y - 5$.	When $y = 12$, $3y - 5 = 31$ and $y + 2(y - 3) = 12 + 18 = 30$
28	**a** $2g - 16$ **b** $g = 18.5$ **c** $g = 8$	$2 + 5g - 3(g + 6) = 2 + 5g - 3g - 18 = 2g - 16$ $2g - 16 = 21 \rightarrow 2g = 37 \rightarrow g = 18.5$ $2 + 5g = 3(g + 6) \rightarrow 2 + 5g - 3(g + 6) = 0 \rightarrow 2g - 16 = 0 \rightarrow g = 8$
29	**a** 1, 4, 9, 16 **b** $14^2 = 196$ **c** 7 **d** 29	 It is $14 + 15$.
30	$(4, -5)$, $(-2, -5)$, $(-2, 1)$	Draw a diagram to show this, if it helps.
31	**a** $y = \frac{1}{2}x + 3$ **b** $y = x^2 - 4x - 3$	It has the same gradient as the blue line and the y-intercept is 3. All the points are translated down 5 units.
32	**a** $x = 11$ **b** $x > 11$	
33	**a** 68 **b** 5	$32 \times 2 + 4 = 68$ The 2nd is $(32 - 4) \div 2 = 14$; the first is $(14 - 4) \div 2 = 5$.
34	**a** 33	$3a + 3b = 20 + 13 = 33$

	b 11	$a + b = 33 \div 3 = 11$
	c 9	$2a + b = 20$ and $a + b = 11$ so subtract to get $a = 9$.
35	**a** Choose some points and show that their coordinates satisfy the equation. For example, if you choose (0, 12) you get $(3 \times 0) + (2 \times 12) = 24$. Two points are sufficient. **b** Rearrange to get $2y = -3x + 24$, then divide by 2.	

3 Ratio, proportion and rates of change

1	£1.45	200 ml cost £0.87 ÷ 3 = £0.29 The cost of a litre is £0.29 × 5 = £1.45
2	**a** 2 : 1 **b** $\frac{2}{3}$ **c** 50%	24 : 12 = 2 : 1
3	£15.73	£4.31 ÷ 0.274 = £15.73
4	7 cm	AC = 15 cm and AD = 8 cm so CD = 7 cm
5	£102.57	
6	**a** 2 : 5 **b** $\frac{1}{5}$ **c** 12.5% **d** Sugar 200 g and butter 100 g	50 : 125 = 2 : 5 $\frac{25}{125} = \frac{1}{5}$ 25 g out of 200 g is 12.5%. 0.5 kg = 500 g. The mass is multiplied by 4.
7	**a** 3.2 m = 320 cm **b** 0.76 m = 760 mm or 0.76 km = 760 m **c** 48 mm = 4.8 cm **d** 0.9 km = 90 000 cm	
8	**a** 22.7% **b** 5 : 4 **c** 4.08 g or 4.1 g	$\frac{6.8}{30} \times 100 = 22.7\%$ 8.5 : 6.8 = 5 : 4 This is a proportion question. 18 ÷ 30 = 0.6; 0.6 × 6.8 = 4.08 or 4.1 g.
9	**a** **b** Top × side is always the same value (144) **c** 12 cm	The equation is $xy = 144$. Because 12 × 12 = 144.
10	There are several ways to do this. Two are shown opposite. Either method shows the large bag is better value.	Method 1: Find the number of grams for 1p. Small bag 250 ÷ 95 = 2.63… g; large bag 450 ÷ 153 = 2.94… Method 2: Find the cost of 100g. Small bag 95p ÷ 2.5 = 38p; large bag 153p ÷ 4.5 = 34p.

11	3%	Her friend pays £6 each month. This is 3% of £200.
12	**a** Multiply both fractions by 4 to get 2 and 3. **b** $\frac{1}{3} : \frac{4}{9} = \left(\frac{1}{3} \times 9\right) : \left(\frac{4}{9} \times 9\right) = 3 : 4;$ $\frac{1}{2} : \frac{2}{3} = \left(\frac{1}{2} \times 6\right) : \left(\frac{2}{3} \times 6\right) = 3 : 4$	
13	20	20 drinks. 250 ml of drink needs 50 ml of squash. One litre = 1000 ml and $1000 \div 50 = 20$
14	172 yen to the pound	$8600 \div 50 = 172$
15	True. 1000 000 seconds = $1000\,000 \div 60 \div 60 \div 24$ hours = 11.57… which is less than 14 days.	
16	Kim £27.20, Jasmine £40.80	Use the fact that the ratio is 3 : 2.
17	**a** 0.568 litres **b** 1.76 pints	This is $4 \div 2.272$.
18	**a** Shade any 15 squares. **b** 5 : 3 **c** 3 squares	Then the ratio is 18 : 6 = 3 : 1.
19	**a** 41.7 metres/minute **b** 24 minutes **c** 4 minutes longer	$20 \times 25 \div 12 = 41.7$ metres/minute Amir takes 16 minutes.
20	**a** 2 : 5 **b** 41.7% **c** It is for the small and medium because $600 \div 250 = 1.56 \div 0.65 = 2.4$. It is not for the small and large bag because $1500 \div 250 = 6$ but $3.70 \div 0.65 = 5.69…$ and not 6	$600 : 1500 = 2 : 5$ $\frac{250}{600} \times 100 = 41.7\%$
21	Copper has greater density. Tin is $19.9 \div 2.6 = 7.7$ g/cm^3 and copper is $37.0 \div 4.1 = 9.0$ g/cm^3	
22	**a** The graph is a straight line through the origin. **b** 154 lb	Use the fact that 20 kg = 44 lb.
23	**a** 1550 m **b** 1023 m **c** 16% or 15.6%	One method is to work out $930 \div 6 \times 10$. One method is to multiply 930 by 1.1 to increase it by 10%. Use the fact that $1075 \div 930 = 1.1559…$
24	**a** The graph is a straight line through the origin. **b** 10 kg **c** 80 g/m^2. One sheet is $2500 \div 500 = 5$ g and $5 \times 16 = 80$	From the graph, 500 sheets have a mass of 2.5 kg.
25	**a** 11.5% **b** 435 g	Mass and percentage are proportional. Answer is $\frac{50}{30} \times 6.9$. It is $\frac{100}{6.9} \times 30$.
26	**a** 5 m/s **b** 18 km/h **c** 11.25 miles/hour	The graph shows that 5 m/s is 18 km/h.
27	A is true because $5.0 \times 1.1 = 5.5$ B is false because $5.5 \times 1.1 = 6.05$ and not 6.0. C is true because $5.0 \times 1.2 = 6.0$.	An alternative method is to work out the percentage increase each time. So for A the increase is 0.5 kg which is $\frac{0.5}{5.0} \times 100 = 10\%$, and so on.

28	**a** 45%	The initial split is £48 − £32 and then it changes to £44 − £36.
	b $\frac{2}{3}$	The split now is 24 − 16 and 16 is $\frac{2}{3}$ of 24.
29	**a** 13 : 7	65 : 35 = 13 : 7
	b 7 : 20	35 : 100 = 7 : 20
	c $\frac{7}{13}$	
	d 646	One method is to calculate $420 \div 65 \times 100 =$ 646.15… Round this to a whole number of people.
30	**a** €60 or $80	
	b €36	You can change it to pounds first: $48 = £30 = €36.
	c No. $120 = £75 = €90	
31	24% because $0.6 \times 40\% = 24\%$	
32	1 : 12	Because there are 4 women to every man and 3 children to every woman. $4 \times 3 = 12$.
33	350 m	The calculation is $21 \times 1000 \div 60$.
34	**a** Choose some points such as (10,36), (20,18), (30,12), (40,9) and show that $x \times y$ is always the same value (360).	
	b £3.60	
35	**a** £69	
	b 15%	Because $69 \div 60 = 1.15$.
	c £425.50	This is $£370 \times 1.15$.

4 Geometry and measures

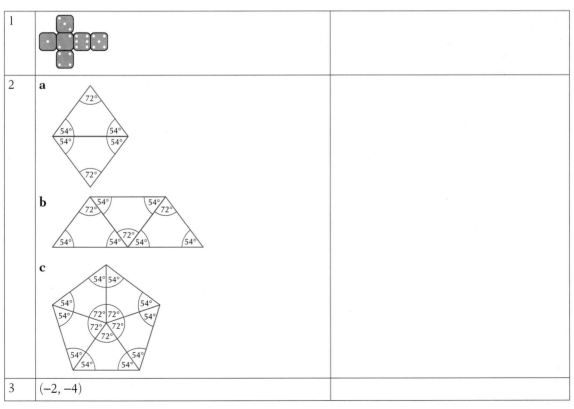

1		
2	**a**	
	b	
	c	
3	(−2, −4)	

4	Phil is not correct. The original area is $\frac{1}{2} \times 4 \times 3 = 6$ cm^2. The enlarged area is $\frac{1}{2} \times 8 \times 6 = 24$ cm^2 which is four times larger.	
5	A rotation of 180° about the bottom left-hand corner; a translation three cm to the right and 3 cm down; a reflection in the diagonal line of symmetry	
6	The areas of the white triangles are 4.5, 9 and 9 cm^2. The area of the red triangle is $36 - (4.5 + 9 + 9) = 13.5$ cm^2	
7	**a** Pyramid and triangular prism **b** Cube and cuboid **c** Triangular prism	
8	She can work out the size of angle EAD $= 180° - a°$. If this is the same as $b°$ they are corresponding angles and the lines are parallel.	
9	$AC = \sqrt{16^2 + 8^2} = \sqrt{256 + 64} = \sqrt{320} = 17.88...$	
10	**a, b** The area is 21 cm^2 so possible sizes of the rectangles are 3 cm by 7 cm or 2 cm by 10.5 cm or 4 cm by 5.25 cm. **c** A parallelogram where the base × the height is 21. One possibility is shown opposite.	There are others.
	d, e Draw triangles where the base × the height is 42 cm^2. Two examples are given opposite. **f** Yes. The area of the circle is $\pi \times 2.6^2 = 21.2$ cm^2.	
11	6 cm	The volume of the cuboid is $8 \times 9 \times 3 = 216$ cm^3. The side of the cube is $\sqrt[3]{216}$ cm.
12	**a** 72 + 20 = 92 42 + 50 = 92 **b** 57 + 35 = 92	
13	**a, b** A reflection in any vertical line followed by a translation can map A onto B. One example is a reflection in the vertical line through the bottom right-hand corner followed by a translation of 2 to the right and 4 up.	There are others.

14	**a** 12.25 cm^2	$\frac{1}{2} \times 7 \times 3.5 = 12.25$ cm^2
	b 3.5 cm^2	$\frac{1}{2} \times 2 \times 3.5 = 3.5$ cm^2
15	**a** Yes	This question is about Pythagoras' theorem.
		$8^2 + 15^2 = 289 = 17^2$
	b Yes	$8^2 + 6^2 = 100 = 10^2$
	c No	$8^2 + 9^2 = 145$ and $12^2 = 144$
16	24 cm^3	The easiest method is to use the fact that the original volume is $3^3 = 27$ cm^3 and 3 cm^3 is removed.
17	Each side of the square is 6 cm. The area is $6 \times 6 = 36$ cm^2. Each side of the triangle is 8 cm. The height must be less than 8 cm. The area is less than $\frac{1}{2} \times 8 \times 8 = 32$ cm^2 so it is less than the area of the square.	
18	Method 1: The exterior angle is $180 - 108 = 72°$. All the exterior angles add up to $360°$ and $360° \div 72° = 5$ so there are 5 sides. Method 2: The angle one side subtends at the centre is $180° - (2 \times 54°) = 72°$. The number of sides is $360° \div 72° = 5$. 	There are several ways to show this.
19	40 cm	First find the hypotenuse $= \sqrt{8^2 + 15^2} = 17$ Then add the three sides.
20	**a** All are 30 cm	$2(3 + 12) = 30$; $2(4 + 11) = 30$; $2(6 + 9) = 30$; $2(7 + 8) = 30$
	b Areas are 36, 44, 54 and 56 cm^2.	
	c Possible answers are 2 cm by 13 cm, or 1 cm by 14 cm, or 2.5 cm by 12.5 cm.	
	d 56.25 cm^2	Each side is $30 \div 4 = 7.5$ cm.
21	Method 1: If the diameter is 2.6 cm, the radius is 1.3 cm and the area is $\pi \times 1.3^2 = 5.3$ cm^2 (to one decimal place). Method 2: If the area is 5.3 cm^2 then $\pi r^2 = 5.3 \rightarrow r^2 = \dfrac{5.3}{\pi} \rightarrow r = \sqrt{\dfrac{5.3}{\pi}} = 1.3$ cm (to one decimal place).	
22	**a** 75°, 105° and 105°	The opposite angle is also 75°. The other two are equal and add up to $360° - (2 \times 75°)$.
	b Either 38° and 104° or 71° and 71°	There are two possible answers, depending on whether there are two 38° angles or just one. You must give both possibilities.

	c Either 60° and 110° or 130° and 40°.	Again you must give both possible answers. There must be two equal angles in a kite. They could be 60° or 130°. The four angles must add up to 360°.
23	**a** The perimeters are 16 cm and 14 cm. **b** This square has a perimeter of 16 cm. 	
24	220 cm (to the nearest whole number)	This distance is the circumference of the wheel.
25	Both are wrong. The cylinders have the same volume. The volume of A is $\pi \times 6^2 \times 3 = 108\pi$ cm^3. The volume of B is $\pi \times 3^2 \times 12 = 108\pi$ cm^3.	

5 Probability

1	**a** $\dfrac{2}{7}$ **b** $\dfrac{2}{7}$ **c** 0	3 and 6 are the multiples of 3. No card fits that description.
2	0.6	Cloudy and dry has a probability of 0.1 and all three probabilities add up to 1.
3	16 red sweets because $\frac{4}{5}$ of 80 is 64. $\frac{4}{5}$ of the sweets are not red.	
4	**a** <table><tr><td>2p</td><td>10p</td><td>20p</td></tr><tr><td>H</td><td>H</td><td>H</td></tr><tr><td>H</td><td>H</td><td>T</td></tr><tr><td>H</td><td>T</td><td>H</td></tr><tr><td>H</td><td>T</td><td>T</td></tr><tr><td>T</td><td>H</td><td>H</td></tr><tr><td>T</td><td>H</td><td>T</td></tr><tr><td>T</td><td>T</td><td>H</td></tr><tr><td>T</td><td>T</td><td>T</td></tr></table> **b** $\dfrac{7}{8}$ **c** $\dfrac{1}{2}$	It is best to list them systematically to make sure you find them all. That means 2 or 3 heads.
5	60%	The probabilities of red, blue and miss must add up to 1.
6	**a** True. The probability of an odd number is $\frac{1}{4}$ so a quarter of the cards are odd. **b** False. 98 is a two-digit number and it could be on one of the cards. **c** False. $\frac{3}{4}$ of the cards are even and $\frac{3}{4}$ of 50 is not a whole number.	

7	14%	It is $100 - (32 + 26)$ divided by 3.
8	**a** Green **b** Orange and grey **c** Yellow **d** $\frac{3}{4}$	The angle is 30° which is $\frac{1}{12}$ of 360°. The pink sector is $\frac{1}{4}$ of the circle.
9	$\frac{1}{10}$ or 0.1 or 10%	The ratio of the two probabilities is 1 : 9 and they add up to one.
10	No. The probability is still $\frac{1}{6}$. It does not change because she has thrown two 4s.	

6 Statistics

1	The bar chart should have a scale that starts at zero and bars of equal width. A possible bar chart is shown opposite.	 Destinations of travellers
2	33	That is $10 + 15 + 8$.
3	**a** E **b** You can only find the mean for numbers, not for letters. **c** 51° **d** There are too many sectors. You cannot show missing letters.	
4	**a** $\frac{1}{3}$ **b** Medians are 7 for boys and 8 for girls **c** Boys because Monday and Friday sectors are equal sizes.	It is 5 out of 15. You could find another reason.
5	**a** $\frac{1}{3}$ **b** $(8 \times 1) + (4 \times 2) + (15 \times 3) + (9 \times 4) + (4 \times 5) + (2 \times 6) + (1 \times 7) = 136$	This is $\frac{16}{48}$ simplified.
6	**a** Friday, with a range of $35 - 14 = 21$. **b** Large size. Monday is smallest and Thursday the largest. **c** Medium	 Because this has the largest total.
7	**a** Cannot tell **b** False **c** False **d** Cannot tell, we do not know the numbers **e** True	You do not know the numbers. Alphaville is Under 21, Betaton is 45 to 64. Alphaville is in the class 21 to 44, Betaton is in the class 45 to 64. You do not know the numbers. The 45 to 64 sector is smaller in Alphaville.
8	**a** The total is $3.6 \times 15 = 54$ **b** The total must be a whole number and $3.1 \times 15 = 46.5$ is not.	

9	**a** A pie chart shows what fraction of the whole population is 16, 17, 18 or 19 and makes it easy to estimate the fraction of the whole. **b** 138°	$\frac{23}{60} \times 360° = 138°$
10	6 goals	The total of the first five is $3 \times 5 = 15$. The total for the six is $3.5 \times 6 = 21$.

7 Mixed questions

1	£45	Look at the difference between west and north or south.
2	**a** All sides 8 cm **b** There are four possibilities: 7, 7, 10 or 9, 9, 6 or 10, 10, 4 or 11, 11, 2. **c** There are six possibilities: 3, 10, 11 or 4, 9, 11 or 5, 8, 11 or 5, 9, 10, or 6, 7, 11 or 6, 8, 10.	
3	The pattern has 15 squares. The tile has 2 squares and can only make patterns with an even number of squares.	You could also use the fact that there are different numbers of black and white squares.
4	**a** Amsterdam and Brussels **b** 32% **c** 131 miles	This is $\frac{210}{656} \times 100$ This is $210 \div 8 \times 5$.
5	**a** $\frac{5}{8}$ **b** $\frac{5}{8}$ of 40 is a whole number, $\frac{5}{8}$ of 50 is not. **c** 20%	This is $\frac{1}{5}$.
6	320	The simplest method is $(300 \times 240) \div (15 \times 15)$.
7	**a** 28% **b** 31–40 **c** 100.8° **d** 88% **e** 5100	This is 28% of 360°.
8	**a** 50 **b** There are 16 doses. **c** 75% **d** 1 : 6	$1000 \div 20 = 50$ $250 \div 15 = 16 \text{ r } 10$ This is 10 : 60 simplified.
9	48 cm²	The original size is 8 cm by 6 cm.
10	**a** $\frac{1}{3}$ **b** 5 : 3 **c** 16.7% **d** 26.7%	There are 10 primes. This is 10 : 6 simplified. There are 5 out of 30 numbers. There are 8 factors.
11	**a** 20 **b** $\frac{7}{12}$ **c** 30	$5 \div \frac{1}{6} = 5 \times 6 = 30$

12	**a** 72	
	b 26.7%	This is 8 out of 30.
	c 2 : 3	This is 8 : 12 simplified.
13	**a** 6 posts and 10 bars	
	b 27 m	
	c No. For example, you need 2 posts for 3 metres and 3 posts for 6 metres. The length has doubled but the number of posts has not.	
	d Yes. The number of bars is always $\frac{2}{3}$ of the length in metres.	
14	**a** 72	This is $1 + 2 + 3 + 5 + 6 + 10 + 15 + 30$.
	b $30 = 2 \times 3 \times 5$	
	c $30 = 2 + 5 + 23$ or $2 + 11 + 17$	
15	**a** 4.7	
	b 3.4	
	c 1.7	This is the x-coordinate of the point where the lines cross.
16	**a** 0.4 s	This is $10 \div 25$.
	b 400 s	
	c 31.25 m/s	
17	$2x + 5 = 5(x - 4)$	
	$2x + 5 = 5x - 20$	
	$5 = 3x - 20$	
	$25 = 3x$	
	$x = 8\frac{1}{3}$	
18	**a** $2(2a + 13)$	Or an equivalent expression such as $4a + 26$.
	b $40 + 8a$	A vertical line will divide the shape into two rectangles.
19	**a** The angles of the triangle are 60°. The marked angles are vertically opposite the angles of the triangle so they are 60° too.	
	b The sum of the angles of the triangle is 180°. The angles marked are the same as the angles of the triangle so they too have a sum of 180°.	
20	**a** $(3 + 5 + 8 + x + x) \div 5 = 9$ so $16 + 2x = 9 \times 5 = 45$	
	b $2x = 29 \rightarrow x = 14.5$	
21	**a** She should subtract the squares. She has not squared the numbers correctly.	
	b Here is a correct version.	
	$x^2 = 12^2 - 9^2$	
	$\quad = 144 - 81$	
	$\quad = 63$	
	$x = \sqrt{63} = 7.9$	
22	**a** $\sqrt{2 \times 8} = \sqrt{16} = 4$	
	b $\sqrt{8 \times 50} = \sqrt{400} = 20$	
	c If the number is N, $\sqrt{3N} = 9 \rightarrow 3N = 81 \rightarrow N = 27$	
	d The product is 100. Possible numbers are 1 and 100; 2 and 50; 4 and 25; 5 and 20; 10 and 10	

23	**a** $x + 12 - x + x + 12 - x = 24$ **b** $x(12 - x) = 12x - x^2$ **c** Either 3.6 or 8.4 **d** The graph does not go as high as 40.	Or an equivalent explanation.
24	$\frac{1}{2} \times 26 \times CD = 143$ so $CD = 11$ cm	
25	50	The difference between terms is $(80 - 68) \div 2 = 6$
26	**a** 31.4 cm **b** 21.46 cm^2	The diameter of the circle is 10 cm. The perimeter is the shape is the circumference of the circle. The area of the shape is the difference between the area of the square and the area of the circle.
27	**a** Choose two points on each line. For example: (1, 2) and (5, 4) satisfy $2y = x + 3$ so this is line A; (0, 7) and (2, 3) satisfy $2x + y = 7$ so this is line B **b** A is −2 and B is $\frac{1}{2}$ **c.** Where the lines cross, $x = 2.2$ and $y = 2.6$	